W0036592

Learning About Your Genes

A Primer for Non-Biologists

Other World Scientific Titles by the Author

New-Opathies: An Emerging Molecular Reclassification of Human Disease
ISBN: 978-981-4355-68-1

A Biography of Paul Berg: The Recombinant DNA Controversy Revisited
ISBN: 978-981-4569-03-3
ISBN: 978-981-4569-04-0 (pbk)

Emperor of Enzymes: A Biography of Arthur Kornberg, Biochemist and Nobel Laureate
ISBN: 978-981-4699-80-8
ISBN: 978-981-4699-81-5 (pbk)

Learning About Your Genes

A Primer for Non-Biologists

Errol C. Friedberg, M.D.
University of Texas Southwestern Medical Center at Dallas, USA

World Scientific

NEW JERSEY · LONDON · SINGAPORE · BEIJING · SHANGHAI · HONG KONG · TAIPEI · CHENNAI · TOKYO

Published by

World Scientific Publishing Co. Pte. Ltd.

5 Toh Tuck Link, Singapore 596224

USA office: 27 Warren Street, Suite 401-402, Hackensack, NJ 07601

UK office: 57 Shelton Street, Covent Garden, London WC2H 9HE

Library of Congress Cataloging-in-Publication Data

Names: Friedberg, Errol C., author.

Title: Learning about your genes : a primer for non-biologists / Errol C. Friedberg, University of Texas Southwestern Medical Center at Dallas, USA.

Description: New Jersey : World Scientific, [2018] | Includes index.

Identifiers: LCCN 2018037039| ISBN 9789813272613 (hardcover : alk. paper) | ISBN 9813272619 (hardcover : alk. paper)

Subjects: LCSH: Genes--Popular works. | DNA--Popular works.

Classification: LCC QH447 .F7535 2018 | DDC 572.8/6--dc23

LC record available at https://lccn.loc.gov/2018037039

British Library Cataloguing-in-Publication Data

A catalogue record for this book is available from the British Library.

First published 2019 (Hardcover)

Reprinted 2019 (in paperback edition)

ISBN 978-981-120-829-4 (pbk)

For any available supplementary material, please visit
https://www.worldscientific.com/worldscibooks/10.1142/11053#t=suppl

Information is Knowledge
Knowledge is Power

Dedicated to Siddhartha Mukherjee in appreciation of the inspiration his books have ignited in my own literary efforts

Contents

About the Author

Errol Clive Friedberg, now retired, was a biologist and historian of science in the Department of Pathology at Stanford University and subsequently the University of Texas Southwestern Medical Center. He studied medicine at the University of Witwatersrand in South Africa and subsequently received postdoctoral training in biochemistry and pathology at Case Western Reserve University before joining the faculty at Stanford University and subsequently the University of Texas Southwestern Medical Center at Dallas.

Friedberg has written several editions of *DNA Repair and Mutagenesis*, published by ASM Press, a number with multiple co-authors, and has published several volumes on aspects of the history of molecular biology, including:

Correcting the Blueprint of Life — An Historical Account of the Discovery of DNA Repair Mechanisms

The Writing Life of James D. Watson

From Rags to Riches — The Phenomenal Rise of the University of Texas Southwestern Medical Center at Dallas

Sydney Brenner: A Biography

A Biography of Paul Berg — The Recombinant DNA Controversy Revisited

Emperor of Enzymes — A Biography of Arthur Kornberg — Biochemist and Nobel Laureate

Friedberg has contributed over 400 papers to the scientific literature, mainly on the topic of DNA repair and is Founding Editor-in-Chief of the scientific journal *DNA Repair*. He has received several awards, including the Rous Whipple Award from the American Society for Investigative Pathology and the Lila Gruber Award for Cancer Research, and is an Honorary Fellow of the Royal Society of South Africa.

Acknowledgments

Writing **UNDERSTANDING YOUR GENES** was inspired by my desire to write a book about a biological topic for a reading audience that is not biologists. I freely admit that having published or co-published a number of books, including the biographies of four Nobel Laureates, and having served as editor of several others, all of which were emphatically written for the scientific community, primarily biologists who work(ed) in the vast world of molecular biology and biochemistry, I found this book the most challenging, and ultimately the most rewarding literary undertaking of my writing career (hobby is a more appropriate word, since I have always identified my career as that of a working biologist).

I owe heartfelt thanks to a cadre of individuals who read multiple drafts of the book and offered cogent advice, notably my wife Rhonda, Laurence Benater, Larry Beaton, Thomas Bonura, Chris Donges and Desmond Levin. I also thank graphic artist Mark Smith from the University of Texas Southwestern Medical Center at Dallas for the drawing in Fig. 3-2.

Last but not least I wish to acknowledge the sterling efforts of Sook Cheng Lim and her colleagues at World Scientific Publishing for their efforts in bringing the published book to reality.

1 Introduction

While once chatting with one of my grandchildren I was faced with a probing question. "Grandpa" he asked, "where did you come from? Not just your father and mother and your grandparents, but before them?" Surprised, but gratified that one of my grandchildren was interested in his genealogy I informed him that while I had no immediate answer to his question, I'd try my best to find one.

Browsing on the Internet, I was intrigued to discover resources such as *MyheritageDNA, KinCore.org and AncestryDNA*; entities that offer genealogical searches based on examination of one's DNA — for a fee of course. The last-mentioned organization touts: *Discover the family story your DNA can tell. Uncover your ethnic mix, discover distant relatives, and find new details about your unique family history with a simple DNA test.*

Then the thought struck me: "How much does the average non-biologist know about DNA and genes in the first place?" Further web surfing about these topics was not especially encouraging. Online information about DNA and genes is scattered, incomplete and all too often presented at a level that is likely beyond the comprehension of most (if not all) non-biologists.

All life forms on earth ranging from the tiniest bacteria to elephants, and even plants, possess **genes**. Genes determine the fact that you may have the same color eyes and/or hair as your mother or father and/or some of your children, and may resemble them in other physical characteristics and even non-physical attributes.

Most people are aware that all living things are endowed with **genes** and that these entities determine much of what you are and how you function as a living organism. But unless you have had a course in biology or genetics it's entirely possible that you know very little if anything about genes and the substance they are made of, called **DNA**, an abbreviation for a complex biological compound that we'll consider in more detail in a later chapter. This book is intended to fill that gap in your general knowledge by informing you what genes are, where in your body they are located, how they function, the dire consequences for your health and sometimes for your offspring if they fail to function normally, how they can be damaged and sometimes repaired — and much more.

Like most intellectual disciplines, biology is a huge field with a vast vocabulary, making it difficult (if not impossible) for non-biologists to understand much (if anything) about genes. However, in my considered opinion there is no *a priori* reason why much of the vocabulary of biology cannot be translated into plain English that any reasonably attentive reader can comprehend. This book was written to achieve that goal. Accordingly, throughout the book I have strived to convey biological concepts and explanations in terms that non-biologists can readily comprehend; essentially to inform you about genes in plain English.

The study of DNA and genes is part of an intellectual discipline called **genomics**, a subject that deals with the structure and function of genes. This is formally distinct from the more familiar subject that you may have heard or read about called **genetics**, a topic that primarily focuses on the study of heredity; how characteristics of living beings are inherited from one generation to the next. This distinction notwithstanding, genomics and genetics are closely related subjects, both words being derived from the Greek *genno*, meaning "to give birth," and in the course of reading this book you will learn that the disciplines of genomics and genetics overlap to a significant extent.

* * *

Many of the discoveries in genomics took place over a period of about 35 years, beginning in the 1940s. Another 15 years or more were required before new and more sophisticated technologies would permit the isolation and characterization of individual genes, including those of complex organisms like you and me. Accordingly, the period between the early 1940s and the end of the 1980s constituted a golden age in the world of genomics. These impressive gains notwithstanding, much remains to be learned about our genes.

As a professor who introduced several generations of students to the mysteries and wonders of both biology and medicine, I found it useful to present biology in a historical context; a framework that traces the emergence and elaboration of new knowledge over time, thereby offering a deeper understanding and appreciation of how the disciplines of biology and medicine evolved. Accordingly, *Learning About Your Genes* is presented with a chronological perspective, a perspective that is intended to provide you with a sense of how our knowledge about genes evolved over time, beginning well before the word "gene" was even invented.

The book will also introduce you to some of the many scientists whose fertile ideas and ingenious experiments in the laboratory led to our current understanding about genes. The book includes explanations of several groundbreaking experiments relevant to our comprehension of genes and their function; explanations and insights that will provide you with a sense of how thoughtfully conceived and informative experiments in genomics are executed in the research laboratory. The book also contains a comprehensive glossary that might help the reader understand and remember new biological terms and their meaning. Regardless of my efforts to explain the attributes of genes in simple language, if you stumble along the way feel free to contact me at errolfriedberg@gmail.com. I'm always available to help your comprehension of genes.

2 A Brief History of the Discovery of Genes

The first documentation of the existence of genes came from studies carried out about 160 years ago by **Gregor Johann Mendel**, a monk at St. Thomas Abbey in the city of Brono, Czechoslovakia. Mendel was born in 1822 with the Christian name Johann. The more familiar name Gregor was added when he joined St. Thomas Monastery years later.

Mendel was the only son of Anton and Rosine Mendel. He had two sisters and the family lived on and worked a farm they had owned for generations. As a child Mendel toiled in the garden and also studied beekeeping, activities that cultivated in him a progressively deepening interest in the biological sciences.

* * *

Most textbooks on genetics render a mere glimpse of Mendel's early life, leading one to wonder how a friar working in an abbey could have gained sufficient knowledge and expertise to carry out genetic experiments. But in fact Mendel received a stellar formal education. At the age of eleven, a local schoolmaster noted his scholarly aptitude and suggested to his parents that he be sent to secondary school to continue his education. The move was a financial strain on his family and was often a difficult experience for Mendel himself, who was prone to bouts of depression for much of his life. Regardless, he excelled in his studies and in 1840 he graduated from secondary school with honors.

Mendel subsequently enrolled in a two-year program at the University of Olmütz in Brno, where he again distinguished himself academically. He also tutored students in his spare time to augment his meager financial

Fig. 2-1. Gregor Mendel.

status. When he joined the Faculty of Philosophy at the University of Olmütz, the head of the Department of Natural History and Agriculture was conducting research on hereditary **traits** (characteristics) in plants and animals, notably sheep, and it's likely that he influenced Mendel's interest in inheritance. Despite suffering from bouts of depression that more than once caused him to temporarily abandon his studies, Mendel graduated from the program in 1843. That same year, against the wishes of his father, who expected him to take over the family farm, Mendel began studying to be a monk. He joined the Augustinian order at the St. Thomas Abbey in Brno, where he was given the familiar name Gregor.

As a student at the university, Mendel had struggled financially to pay for his studies and he became a friar at St. Thomas's Abbey primarily to take advantage of the free education offered at the abbey. The monastery was then a cultural center for the region and Mendel was frequently exposed to the research and teaching of its members. He also gained access to the monastery's extensive library and experimental facilities. Mendel eventually became an Augustinian Monk, a religious order in the Roman Catholic Church.

* * *

Those in charge of St. Thomas's Abbey were sufficiently impressed with Mendel's intellectual aptitude that in 1851 they gained him entrance to the University of Vienna at their expense in order to support his continued studies in the sciences. While In Vienna, Mendel studied mathematics and physics under the famous Austrian physicist Christian Doppler, who in 1842 had described the Doppler effect; a physical phenomenon that eventually led to the use of Doppler radar, currently used in weather forecasting. Mendel also studied botany under Franz Unger, an Austrian botanist who had begun using a microscope in his studies, and who was a proponent of a pre-Darwinian version of evolutionary theory. Clearly, Mendel's university education 160-odd years ago was at the cutting edge.

At the university, Mendel exhibited a talent for teaching, though he surprisingly twice failed the teaching certificate examination! He was quiet and shy and perhaps found the oral part of the examination nerve-wracking. Upon completing his studies in Vienna, Mendel returned to the monastery in Brno and was granted a teaching position at a secondary school, a position that he retained for more than a decade. It was during this time that he began the breeding experiments for which he is posthumously famous.

Mendel began his studies on breeding using mice. He bred them in his two-room apartment at the monastery, attempting to discern what sort of offspring would arise when normal-looking mice were mated with those lacking pigmentation (albinos). Would the baby mice have coats marked with traces of each parent, or would one be dominant? Mendel never had an opportunity to answer this question because his bishop, who took offense at the idea of a priest having anything to do with sex, ordered him to discontinue these studies. It was then that Mendel turned to pea plants, pleased that the bishop did not understand that plants also have sex!

* * *

Mendel examined the inheritance of seven traits in pea plants: seed shape (round or wrinkled), pea color (yellow or green), flower color (purple or white), flower position (terminal or axial), plant height (tall or short), pod shape (inflated or constricted) and pod color (yellow or green). He noted whether the offspring of pea plants with round seeds mated to peas plants with wrinkled seeds were round or wrinkled, and the offspring of yellow pea plants mated to green pea plants were yellow or green, and so on for each of the seven traits he had selected for study. Between 1856 and 1863 he is said to have cultivated and examined some 28,000 plants, 40,000 flowers and nearly 400,000 seeds, during what is considered the first well-controlled study in genetics. At the conclusion of this seven year study Mendel speculated that there are **two "factors" for each inherited characteristic and that one factor is inherited from each parent**.

Two years later, Mendel presented the results of his work to the Brno Society for Natural Science. His presentation, entitled *Experiments on Plant Hybridization* was published the following year. While his work was appreciated for its thoroughness, no one grasped its significance. The work was simply too ahead of its time; too contrary to popular beliefs about heredity such as the notion of "blending" inheritance subscribed to by ancient philosophers and natural historians. Hippocrates, for example, considered the founder of medical science, had propounded a theory according to which minute particles from every part of the body entered the "seminal substance" of the parents and by their fusion gave rise to a new individual exhibiting the traits of both of them. But, regardless of the poor reception to his studies, Mendel is quoted as having confidently stated: "My scientific studies have afforded me great gratification; and I am convinced that it will not take long before the whole world acknowledges the results of my work." His conviction was well founded!

* * *

Mendel did little to promote his own work and the few references to his studies indicate that much of it had been misunderstood. Furthermore, his findings were not thought to be generally applicable. In fact, Mendel himself surmised that they only applied to certain types of inherited traits. It was not until decades later, when Mendel's research informed the work of several noted geneticists, botanists and biologists conducting research on heredity, that its significance was more fully appreciated, and his studies began to be referred to as **Mendel's Laws**.

The Law of Segregation: a law that states that each genetic trait is defined by a pair of factors. Parental factors are randomly separated to sex cells, now usually called germ cells — sperm cells in males and ova (eggs) in females — so that sex cells contain only one of the pair of parental factors. When sex cells unite during fertilization of an egg cell by a sperm cell, offspring inherit one genetic form from each parent.

The Law of Independent Assortment: **a law that states that factors for different genetic traits are sorted separately from one another**, i.e., the inheritance of one trait is not dependent on the inheritance of another.

The Law of Dominance: **a law that states that** an organism with alternate forms of a factor (dominant or recessive) will express the form that is dominant.

If the generality of Mendel's laws sound familiar you can surely appreciate how close to the truth he was 160-odd years ago!

* * *

In 1900, three celebrated early geneticists independently duplicated Mendel's experiments and results, allegedly discovering after the fact that both Mendel's data and his general theory had been published in 1866. Questions arose about the validity of the claims that the trio of botanists were not aware of Mendel's previous results, but they soon credited Mendel with priority. Even then, however, his work was sometimes marginalized by respected biologists of the day.

As genetic theory continued to develop, the relevance of Mendel's work fell in and out of favor, but his research and theories are ultimately considered fundamental to any understanding of the field we now call **genetics**, and he is thus deservedly considered the "**father of modern genetics**."

It took more than 30 years before Mendel's work was fully acknowledged — a time when he was no longer alive to celebrate the recognition of his groundbreaking labors. But his legacy has not been lost. Most, if not all, genetics courses in schools, colleges and universities around the world teach his work, and contemporary geneticists often use the term "Mendelian genetics" when discussing our present understanding of genetics. Mendel died in 1884 at the age of 62. While he never called his "inheritance factors" genes, the reality is that through his breeding studies with pea plants, Mendel identified the concept of **hereditary elements that we now call genes!** In a recent book entitled *The Gene — An Intimate History*, author Siddhartha Mukherjee eloquently stated:

> **Behind the epic variance of natural organisms — tall: short; wrinkled; smooth; green; yellow there were corpuscles of hereditary information, moving from one generation to the next. Each unit was *unitary* — distinct, separate, and indelible.**
>
> **Mendel did not give this unit of heredity a name, but he had discovered the most essential features of a gene.**

The word **gene** was first used by a Danish botanist, Wilhelm Johannsen, a professor of botany and plant physiology in Copenhagen, who wrote:

> The word "gene" — expresses — the evident fact that — many characteristics of [an] organism are specified — by means of — determiners which are present in unique, separate, and thereby independent ways — **in short, precisely what we wish to call genes.**

3 Genes are Made of DNA

Most studies of biology ultimately seek an understanding of the biology of humans — of you and me! However, many biological experiments are technically challenging and frequently involve procedures that result in death of the organism under study. For these and other cogent reasons the use of human subjects for many biological studies is forbidden by most if not all agencies that provide financial support for biological research. Consequently, molecular biologists, biochemists, geneticists and other members of the biological research community have for many years explored the biology of organisms that are inexpensive to maintain and house, multiply quickly, and are easy to handle and store in the laboratory. Organisms ranging from simple bacteria to flies to rodents and even primates such as monkeys have over the years served as the mainstay of biological research and are referred to as **model organisms**. Bacteria are frequently used model organisms for the study of genes, particularly a bacterium called *Escherichia coli* (*E. coli* for short), a generally harmless organism present in the intestines of many animals, including humans.

It might strike you as strange that studying simple bacteria (tiny organisms that require a powerful microscope just to view them) has yielded fundamental information about your genes. But the reality is that the structure and function of genes have not changed significantly during the evolution of bacteria to humans. Consequently much of what we know about genes and gene function — and in fact about biochemistry and molecular biology in general, stems from studies using simple bacteria such as *E. coli*. Indeed, a popular aphorism in science attributed to the French molecular biologist Jacques Monod states: "what's true for *E. coli* is true for the elephant."

* * *

Ever since Gregor Mendel's seminal studies were published, scientists conjectured about what his "inheritance factors" (genes) are made of and where they reside in our bodies. In the 1920s **Frederick Griffith**, an English scientist and medical officer in the British Ministry of Health, was classifying different strains of a bacterium called *Streptococcus pneumonia*, a dangerous organism that can cause pneumonia, that in pre-antibiotic times not infrequently led to death. Born in 1879, details of Frederick Griffith's life are not documented. He was a shy and reticent individual whose quiet manner made him a well-liked personality to the few who knew him well.

In his 2016 book *The Gene — An Intimate History*, author Siddhartha Mukherjee wrote:

> **Griffith lived alone, in a nondescript apartment near his lab in London, and in a spare, white modernist cottage that he had built for himself in Brighton. Genes might have moved between organisms, but Griffith could not be forced to travel from his lab to his own lectures. To trick him into giving scientific talks his friends would stuff him into a taxicab and pay a one-way fare to the destination.**

* * *

Griffith wasn't interested in addressing questions about genes. His focus was on whether different strains (types) of *Streptococcus pneumonia* collected from different patients were more likely to cause lethal pneumonia than others. He collected samples of this bacterium from numerous patients with pneumonia in different parts of England and classified them in search of patterns of the disease, using mice as his experimental subjects.

Streptococcus pneumonia exists in two forms called *rough* (R) and *smooth* (S). The S form can cause serious pneumonia, frequently killing infected subjects, and is accordingly referred to as the **virulent form** of *Streptococcus pneumonia*. In contrast, the R form of this bacterium causes less serious illness, sometimes none at all, and is called the **non-virulent form**.

In 1927, Griffith made an observation for which he is posthumously remembered (but not renowned) in scientific circles. He discovered that when the non-virulent (R-type) bacteria were mixed with the virulent (S-type) form that had been inactivated (killed) by heating, **the non-virulent strain became virulent**! Griffith referred to this phenomenon as **transformation** and suggested that the phenomenon was caused by "something" in the non-virulent R-type bacteria that he called a **transforming principle**. But he made no attempt to determine what his transforming principle was made of. Siddhartha Mukherjee wrote:

> **In January 1928, after hesitating for months ("God is in no hurry, so why should I be?"), Griffith published his data in the *Journal of Hygiene* — a scientific journal whose sheer obscurity might have impressed even Mendel. Writing in an abjectly apologetic tone, Griffith seemed genuinely sorry that he had shaken genetics by its roots. His study discussed transformation as a curiosity of microbial biology, but never explicitly mentioned the discovery of a potential chemical basis of heredity. The most important conclusion of the most important biochemical article of the decade was buried like a polite cough, under a mound of dense text.**

* * *

The burning question that emerged from Griffith's experiments concerned the nature of the transforming principle. Aware of Griffith's work that he had first learned about in 1933 (5 years after Griffith published his unremarkable paper in the *Journal of Hygiene*), this question piqued the interest of a physician/scientist named **Oswald Theodor Avery**.

Born in 1877 in Halifax, Canada, Avery received a medical degree from Columbia University College of Physicians and Surgeons in New York City in 1904. In October 1910, the new Rockefeller Institute Hospital admitted its first research participants, opening up a new era of biomedical investigation in which physicians were given the resources and encouragement to engage in fundamental studies of the disease problems they dealt with on the hospital wards. Avery moved to the Rockefeller Institute in 1913, where he focused most of his research for the next 35 years on *Streptococcus pneumonia*, the same organism that his predecessor Frederick Griffith had studied.

Avery was described as something of a loner. Though well liked by his university colleagues, he didn't spend much time socializing with them.

He also traveled infrequently and rarely attended scientific conferences or meetings. A notable exception was his annual summer vacation to Deer Island, Maine, where he indulged in sailing, one of his favorite pastimes. After retirement, Avery moved to Nashville, Tennessee to be near his brother's family, where he cultivated the lifestyle of a retired "country gentleman," taking long walks, gardening and spending time with his family.

In the spring of 1940, Avery confirmed the results of Griffith's experiments. Now 63, an age at which faculty members in many American universities were required to retire, Avery was awarded emeritus status at Rockefeller University, an honor granted by academic institutions to a limited number of distinguished faculty who had reached the age of retirement, but who continued to offer the promise of significant scientific contributions — though without salary support. Emeritus status allowed Avery to continue working at the Rockefeller for a further five years, presumably living off his retirement funds.

When Avery began the studies for which he became renowned, the majority of biologists interested in genes were of the firm opinion that they are made of protein. While at Rockefeller University, Avery, with the help of two junior colleagues, made extracts of bacterial cells in order

Fig. 3-1. Oswald Avery.

to isolate and purify the cellular component(s) of the virulent (S-type) bacteria that caused pneumonia in mice. After much hard work, Avery and his co-workers succeeded in purifying a component of cells that by itself transformed non-virulent bacteria to the virulent form. The component turned out to be DNA, leading Avery to the consideration that when Griffith had transformed bacteria of the non-virulent R type to the virulent S type this resulted from **DNA** containing one or more **genes** that determined virulence to enter R-type cells.

* * *

Most readers have heard of **DNA** and may even know that it is a biological compound that genes are made of. For those who have not, DNA is an abbreviation for the tongue twister **D**eoxyribo**N**ucleic**A**cid (pronounced **DEEOXY-RIBO-NUKLEE-IK-ACID**). This polysyllabic name is based on several chemical properties. In the first instance DNA contains a **sugar** called **ribose.** But, never having tasted DNA, I'm not sure whether it is in fact sweet to the taste. The definition of sugars is based strictly on chemical considerations! Secondly, DNA is located in the **nucleus** of cells **(Fig. 3-2)**, where it exists in tiny threadlike structures called **chromosomes**, about which more is related in Chapter 5. Thirdly, DNA fits the chemical definition of **acids**, but it is not a liquid in the familiar sense that we usually associate with acids. The "**deoxy**" part of the name is also based on chemical nomenclature. Later we'll encounter a related substance without the deoxy component called **ribonucleicacid (RNA)**.

DNA is found in the great majority of the approximately 37.2 trillion cells calculated to exist in the human body. All tissues and organs in your body are composed of **cells** — tiny entities about 100 micrometers in diameter (a micrometer is one millionth of a meter) that can be readily visualized under the microscope. In the event that you are not familiar with human cells, consider a cell resembling a tiny hollow ball. The shell of the ball is called the **plasma membrane (Fig. 3-2)**. The inside of the ball is filled with a liquid called **cytoplasm** that is mainly composed of water, salts, and proteins. Multiple structures are embedded in the cytoplasm **(Fig. 3-2)**. One of these structures, of which there are many in any given cell, are called **mitochondria (Fig. 3-2)**. Biochemical events that generate the energy that our cells require occur in mitochondria. A later chapter

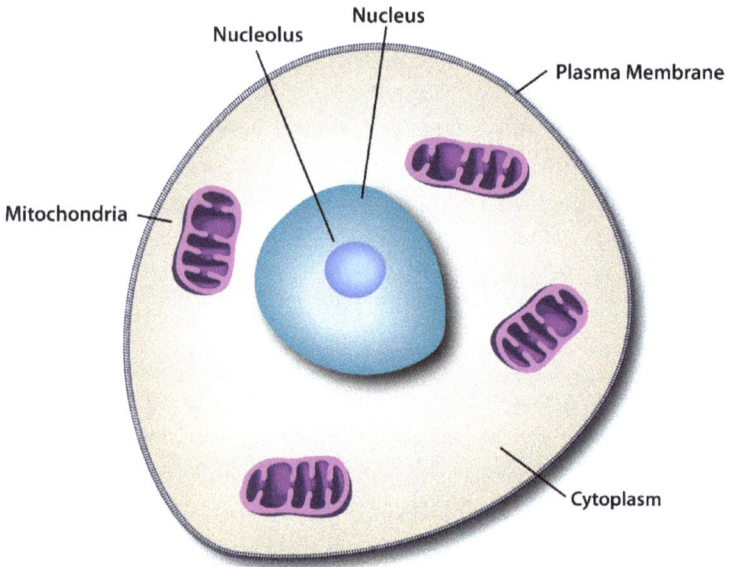

Fig. 3-2. TYPICAL HUMAN CELL. The cellular structures you may want to familiarize yourself with are the **plasma (cell) membrane**, the **cytoplasm,** the **mitochondria** (four of which are shown in this cartoon, though individual cells can contain hundreds), and most importantly, the **nucleus** (where DNA is found), which contains a small structure called the nucleolus that you needn't concern yourself with.

will inform you that much to the surprise of the community of biologists, mitochondria were relatively recently discovered to contain DNA!

* * *

In May 1943, Oswald Avery wrote a letter to his brother Roy, a professor of microbiology at Vanderbilt University, Tennessee, informing him of his belief that Fred Griffith's "transforming factor" was made of DNA. He wrote:

> If we are right, and of course that's not yet proven, then it means that nucleic acids are not merely structurally important, but functionally active substances in determining the biochemical activities and specific characteristics of cells — and that by means of a known chemical substance it is possible to induce predictable and hereditary changes

in cells. This is something that has long been the dream of geneticists. Sounds like a virus — may be a GENE.

One of the fundamental attributes of research in any field is that no matter how initially persuasive a new discovery may be, it must pass rigorous testing and evaluation before gaining unqualified acceptance by the larger scientific community — and the world. Absent this critical vetting process, an enormous amount of time, work, and money can be for naught. Given this reality, one that all scientists must face in the wake of important new discoveries, there was considerable reluctance to accept Avery's contention that genes are made of DNA.

This bias stemmed primarily from the prevailing notion that the hugh variation of biochemical events in cells known to be determined by genes surely required a substance that existed in many forms, such as proteins, of which there are about 20,000 in humans. DNA on the other hand, composed primarily of just four repeated chemical entities called **bases** named **adenine (A)**, **thymine (T)**, **guanine (G)** and **cytosine (C)**, was widely thought of as a compound unsuited for the anticipated role of genes and was sometimes derisively referred to as a "boring" molecule.

It's not essential to remember the names adenine, thymine, guanine and cytosine. But you will find it helpful to recall the letters **A, T, C** and **G**. They will appear multiple times in the book. In addition to the entrenched bias that DNA was a "boring molecule" unsuited to be the stuff that genes are made of, many voiced the opinion that, notwithstanding Avery's heroic efforts to isolate absolutely pure DNA, it was doubtful that his final product was completely free of protein. One influential biologist wrote a letter to a colleague stating "there can be little doubt in the mind of anyone who has purified DNA that traces of protein probably remain in even the best preparations and that as much as 1% or 2% of protein could be present in a preparation of 'pure, protein-free' nucleic acid." Critics argued that even though Avery repeatedly demonstrated that he could detect no protein in the highly purified material that caused transformation, even the tiniest amount of the material that he and his colleagues purified and that worked in their system could contain millions of protein molecules. "Avery wanted to prove a negative; to show that his purified extracts were completely free of proteins, and this was impossible," one historian of science cogently commented.

* * *

But not everyone in the scientific community was skeptical about Avery's discovery. The revered science journal *Nature* described his work in glowing terms and some scientists were in fact highly complementary. In October 1944, the New York Academy of Medicine awarded Avery its Gold Medal. A year later the Royal Society of London graced his experimental achievements with the Copley Medal, and in 1947 Avery was awarded the Albert Lasker Basic Medical Research Award. He was nominated for a Nobel Prize numerous times, but never received that much-coveted honor. Arne Tiselius, a Swedish biochemist who won a Nobel Prize in 1948 is said to have commented that Oswald Avery was the most deserving scientist to not receive a Nobel Prize for his work! And in his book *The Gene — An Intimate History*, Siddhartha Mukherjee commented that Avery was denied the prize "because Einar Hammarsten, an influential Swedish chemist, refused to believe that DNA could carry genetic information."

Following Avery's assertion that genes are made of DNA, an historically important experiment executed by two scientists, **Alfred (Al) Hershey (Fig. 3-3)** and **Martha Chase (Fig. 3-4)** lent much credence to his contention. Al Hershey was born in Owosso, Michigan. He graduated from Michigan State University in 1930 with a B.S. degree and obtained his Ph.D. in 1934. He subsequently accepted a position at the Washington University School of Medicine in Seattle, where he began working on viruses that infect bacteria, called **bacteriophage**, or **phage** for short. Bacteriophage are tiny organisms, much smaller than bacteria, comprised of just DNA with its complement of phage genes, surrounded by a shell of proteins. Over the years, bacteriophage have been an important source of information about genes. Many Ph.D. theses in the biological sciences have come from studies using these bacterial viruses.

Martha Chase was born in 1927 in Cleveland, Ohio. In 1950 she received her bachelor degree from the College of Wooster, a private liberal arts college located in Wooster, Ohio, known for its emphasis on mentored undergraduate research. Chase subsequently worked with

Fig. 3-3. Alfred Hershey.

Fig. 3-4. Martha Chase.

Hershey as a research assistant at the Cold Spring Harbor Laboratory, a highly regarded research facility located near New York City. Though she obtained a Ph.D. degree from the University of Southern California in 1964, a series of personal setbacks during the 1960s ended Martha Chase's career in science. She tragically spent decades suffering from a form of dementia that robbed her of short-term memory, and died in 2003 at the age of 75.

"Her name will always be associated with that experiment, so she is some sort of monument," her longtime colleague and friend Waclaw Szybalski who met Martha Chase when he joined the Cold Spring Harbor Laboratory in 1951, commented. Szybalski attended the first staff presentation of the Hershey–Chase experiment and was so impressed that he invited Chase for dinner and dancing the same evening. "I had an impression that she did not realize what an important piece of work she had done. But I think that I convinced her that evening," Szybalski stated. "Before, she was thinking that she was just an underpaid technician. In fact, experimentally she contributed very much."

In their 1952 collaboration Hershey and Chase carried out an ingenious experiment to support Oswald Avery's contention that genes are made of DNA, not protein. If you are not able to follow the logic of this experiment don't fret! It is one of a few experiments described in the book that are included primarily to demonstrate how investigators in the field of genomics sought experimental evidence in support of ideas and hypotheses that emerged from the fertile minds of the day.

* * *

In 1952, it was known that a substance called **phosphorus** is present in DNA, including that in bacteriophage, but is not found in their protein shell. Conversely, Hershey and Chase were aware that another substance called **sulphur** is found in the protein shell of phage, but not in its DNA. With this knowledge in hand, Hershey and Chase tagged (labeled) phage DNA with radioactive phosphorus, and independently tagged the protein coat of a different group of phage with radioactive sulphur. The two batches of tagged phage were mixed and allowed to infect bacteria.

When phages infect bacteria they shed their protein shell and new copies of DNA carrying genes that encode phage proteins are synthesized.

In this manner new phage are generated. Eventually the phages burst their bacterial hosts and go in search of other bacteria to infect. Hershey and Chase observed that phage that were descendants of those labeled with phosphorus in their DNA still contained this radioactive material. But the descendants of the phage labeled with sulfur in their protein coats were not radioactive. In a nutshell, Hershey and Chase established that it's the **genes** in the DNA of the phage that are important for making new phages, not the proteins. In due course these, and experiments executed by others, removed any doubt that genes are made of DNA. In 1969, Albert Hershey was awarded a Nobel Prize. One assumes that many female biologists have wondered why Martha Chase was not also so honored!

4 The Structure of DNA

In order to understand genes and their role in instructing cells to synthesize (manufacture) proteins, it's important to understand the **structure of DNA**, a structure elucidated in 1953 by an American scientist **James** (known to the scientific world as **Jim) Watson** and an English colleague **Francis Crick**; legendary names you may be familiar with, who were awarded Nobel Prizes in 1962. The elucidation of the structure of DNA by Watson and Crick is considered by many as one of the most momentous and far-reaching discoveries in biology.

Born in Chicago, Illinois in 1928, Jim Watson attained a Bachelor of Science degree in zoology in 1947. As a youth he was an avid bird watcher and might have become an ornithologist. But he ultimately decided that he wanted to learn about genetics. Watson received a fellowship for graduate study at Indiana University, where he encountered and was deeply influenced by several established leaders in genomics and molecular biology of the day.

After receiving a Ph.D. degree in 1950, Watson elected to spend a post-doctoral year in Copenhagen working on the biology of bacteriophage, the tiny viruses that infect bacteria mentioned earlier; work for which he received financial support from a funding agency in the United States as a post-doctoral fellow. In the course of that year, Watson attended a scientific meeting in Naples, Italy. During one of the scientific presentations at that meeting, an English scientist, Maurice Wilkins, showed a slide of an X-ray pattern of DNA generated by a technique called **X-ray crystallography**, a technique used by physical chemists to determine the three-dimensional structure of biological compounds. Watson was entranced with Wilkins's

Fig. 4-1. Francis Crick.

Fig. 4-2. James Watson.

photograph. "The fact that I was unable to interpret it did not bother me," he later wrote. "It was certainly better to imagine myself becoming famous than maturing into a stifled academic who had never risked a thought." Aware that solving the structure of DNA might reveal crucial information about genes, Watson was inspired to switch from the research work he was doing in Copenhagen to confront this challenge.

Funding agencies do not look favorably on scientists switching midstream from research in an approved program in a stipulated laboratory in order to execute different work in a different laboratory. However, after some hankering and with the persuasive assistance of his mentors in America and Copenhagen, the funding agency that was providing financial support for Watson's work granted him permission to switch his research focus. Arrangements were made for him to move to the Cavendish Laboratory in England to learn the rudiments of X-ray crystallography and to pursue structural studies of DNA. At the Cavendish, then a man of just 23, Watson met Francis Crick, 12 years his senior and a brilliant theoretical physicist with experience in X-ray crystallography. After several stutters and starts, Watson and Crick arrived at a model of the structure of DNA in early 1953 that proved to be correct and highly informative about the biology of DNA.

* * *

In 1968, Watson published an enthralling account of this breakthrough in biology in a book entitled *The Double Helix — A Personal Account of the Discovery of the Structure of DNA*. I'll explain where the name *The Double Helix* comes from later in this chapter. If you are interested in the history of this seminal period during which Watson and Crick solved the structure of DNA, the fits and starts involved and the many key personalities at the Cavendish Laboratory and leading scientific venues in London, I urge you to read *The Double Helix*, a book that has sold well over a million copies and has been translated into multiple languages. In 1998 the Modern Library, a renowned American publisher of classic books, chronicles, essential writings and translations, placed *The Double Helix* at #7 on its list of the 100 best non-fiction books of the 20th century. And in 2012 *the Library of Congress named The Double Helix one of the 88 Books That Shaped America*.

A little known feature of the book is that Francis Crick was initially vehemently opposed to the publication of *The Double Helix*, in which Watson revealed many idiosyncrasies of the major scientific personalities involved in the discovery — including Crick! The opening sentence of *The Double Helix* reads: "I have never seen Francis Crick in a modest mood."

In 2003, I was invited by Watson to write a book about his writing life, a labor of love that provided access to his private papers and to several interviews with him. The book, which appeared in 2005 under the title *The Writing Life of James D. Watson*, reveals (among other topics) details of the tussle that transpired between Watson and Crick before Crick reluctantly conceded to Watson's unrelenting intention to publish *The Double Helix*.

Besides Watson and Crick, other competing scientists in London who were independently pursuing the elucidation of the DNA structure included Maurice Wilkins (who shared the 1962 Nobel Prize with Watson and Crick) and Rosalind Franklin, a competitor who Watson once described as a "belligerent, emotional woman unable to interpret her own data." A biography of Rosalind Franklin was rendered by Brenda Maddox, an English historian of science, in a book entitled *Rosalind Franklin — The Dark Lady of DNA*, published in 2003, another book that merits reading.

In his later years, Watson (aged 89 at the time of this writing) was a much talked about figure who was prone to offering opinions on all manner of topics, some of which were not especially flattering. All in all the portrait of a most interesting personality! Consequently, many historians of science (including the author of this book) aspired to document his life. This much-awaited literary plum (and formidable challenge) was awarded to Brenda Maddox. Word of mouth has it that her eagerly anticipated biography of Watson is in the pipeline.

Watson is a prolific writer. In addition to *The Double Helix* he published multiple editions of a major textbook on molecular biology, as well as books entitled *A Passion For DNA*, *DNA — The Secret of Life*, *Avoid Boring People*, and most recently a volume called *DNA — The Story of the Genetic Revolution*.

After returning from his celebrated sojourn in England, Watson joined the faculty of the Department of Biology at Harvard University. In 1976 he moved to the Cold Spring Harbor Laboratory, mentioned earlier. In 1968 Watson assumed the directorship of that facility, greatly expanding its funding for research and elevating the Cold Spring Harbor Laboratory to one of the premier biological research facilities in the world. He now resides at the Cold Spring Harbor Laboratory with his wife Elizabeth (Liz), a writer in her own right who co-authored a book with her husband about the architecture at the Cold Spring Harbor Laboratory entitled *Houses for Science*.

* * *

Francis Harry Compton Crick was born in 1916 in Northampton, England, and was educated at Northampton Grammar School and Mill Hill School, London. He studied physics at University College, London and began research toward a Ph.D. degree that was interrupted by the outbreak of World War II. During the war, Crick worked as a scientist for the British Admiralty. He subsequently attended Cambridge University and obtained a Ph.D. in 1954, a year after his seminal work with Watson on the structure of DNA. Following his triumphant partnership with Watson, Crick switched his research focus to deciphering the **genetic code**, a topic presented later in this book. He subsequently moved to the Salk Institute in La Jolla, California to study neurobiology, the study of the nervous system, principally the brain, an organ that continues to challenge the intellectual talents of numerous investigators — but is yet to reveal its deepest secrets. At one time the Crick family lived in a house named *The Golden Helix*! Francis Crick died in 2004.

In 2016 the *Francis Crick Research Institute* opened its doors in London. At the time of this writing the Institute is planned to have a staff of 1,500 scientists, research technicians and administrators, including 1,250 scientists and an annual budget of over £100 million, making it the biggest single biomedical laboratory in Europe, a fitting memorial to one of the most brilliant investigators in biology of the 20th and 21st centuries.

* * *

Now to the details of the **DNA structure**. DNA molecules consist of two **anti-parallel chains** (a configuration that I'll explain presently), or **strands** as they are called, that lie close together, are of equal length and have a right-handed **helical** (spiral) configuration **(Fig. 4-3, RIGHT PANEL)**, hence the name of Watson's book *The Double Helix*.)

The absolute length of the two DNA strands varies in different genes. Each strand consists of the 4 bases **A**, **T**, **G** and **C** mentioned earlier, which are attached to a sugar-phosphate backbone **(Fig. 4-4)**. The four bases can

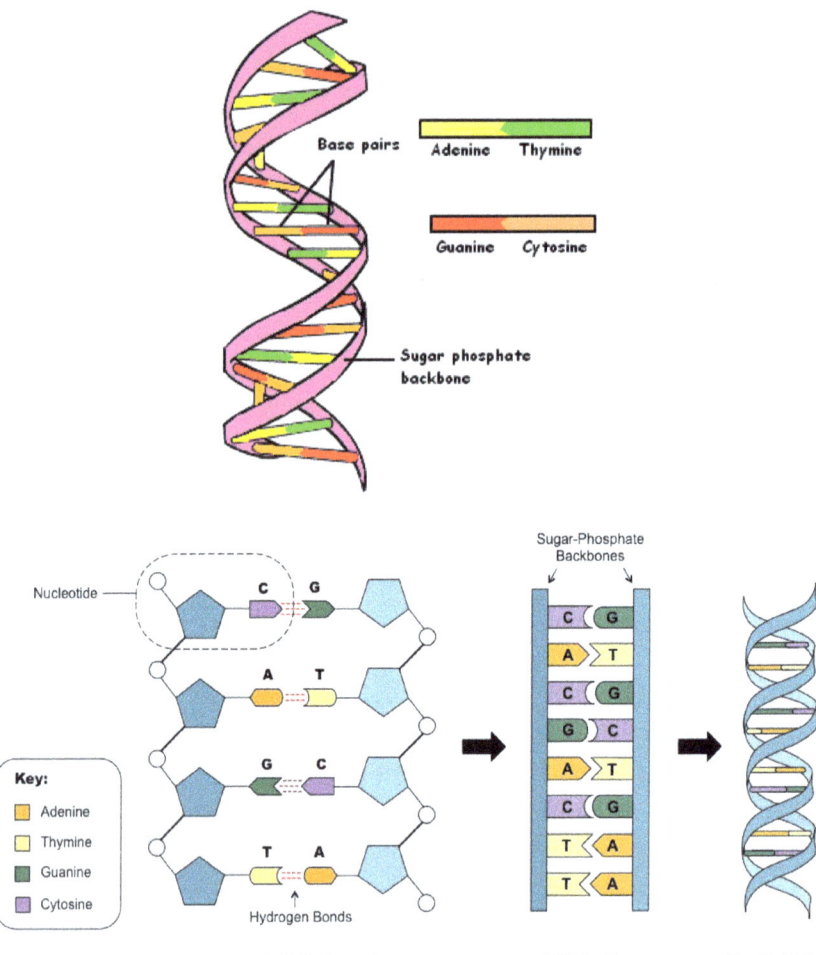

Fig. 4-3. The structure of DNA. Both figures explain the structure of DNA. The diagrams show the two DNA helical strands (the double helix) that comprise a DNA molecule. Note that the base **adenine (A)** is always paired with the base **thymine (T)**, and the base **guanine (G)** is always paired with the base **cytosine (C)**. The bases are chemically linked to a sugar–phosphate backbone (see top figure). The two DNA strands are oriented in an **anti-parallel manner**, such that the "top end" of one DNA strand is aligned with the "bottom end" of the other DNA stand. This is well shown in the bottom figure, in which the pentagonal green structures (that represent the sugar deoxyribose) **point in opposite directions in the two strands**. An easy way of understanding an anti-parallel alignment is to view two dinner forks of the same size side-by-side, but facing in opposite directions (**anti-parallel**). You will observe that the width of the two forks is identical along their entire length. But if the forks face in the same direction (**parallel**) their width is not uniform. The left panel of the bottom figure also shows that in DNA the bases exist as entities called **nucleotides**, composed of a **base** joined to the sugar deoxyribose and a phosphate molecule.

be in any order in a DNA strand. However, a particular order is unique to any given gene. The number and order of the bases in the two strands of DNA are what distinguish one gene from another. **No two genes consist of exactly the same order and number of bases.**

The organization of the two strands in DNA is such that **the base A in one DNA strand is always paired with the base T in the opposite strand, and the base G is always paired with the base C.** If, for example, one of the two DNA strands of a gene contains a nucleotide sequence

$$\textbf{ATCTGTTACGC}$$

the opposite DNA strand will have the sequence

$$\textbf{TAGACAATGCG}$$

A single pair of bases on opposite DNA strands (such as A–T or G–C) is referred to as a **base pair**. **The width of an A–T base pair and a G–C base pair are identical**. If that was not the case, the width of a DNA molecule would vary haphazardly from one base pair to another. Imagine how wobbly a ladder would look (and function) if the steps were not of equal width. **This explains why the base A in one DNA stand is always opposite the base T in the other strand and the base C is always opposite the base G in the other strand.**

You may be wondering how many base pairs (**A–T and C–G**) comprise the entire human genome. The answer is a staggering **3.3 billion**, only about 2% of which directly code for proteins. Of the rest, about 25% make up other genes and their regulatory elements. Believe it or not, in 2018, 65 years after the elucidation of the structure of DNA by Watson and Crick, the function(s) of the remaining bases is unknown. It has been suggested that some of it may be redundant information left over from our evolutionary past.

5 Chromosomes and DNA Replication

The precise number of genes in the human **genome** (a word that describes your entire complement of genes) is not known since they are difficult to count. Presently the number of known human genes that carry an instructional code for making proteins is believed to be somewhere around 20,000. Other genes are known to regulate the genes that specify proteins. But the total number of genes known to instruct cells to make proteins, plus those that are known to regulate this function only occupy about 2% of the DNA in a cell. The remaining stretches of DNA in our cells have no known function at this time! Clearly there is still much to learn about DNA.

As mentioned in an earlier chapter, DNA is present in structures called **chromosomes** that reside in the nuclei of the estimated 37.2 trillion cells in your body. These 37 trillion-odd cells each contain **46 chromosomes organized into 23 pairs (Fig. 5.1)**. The 23 pairs of chromosomes contain two copies of each of your genes; **one** gene that you inherited from your **mother** present in one of the paired chromosomes, and **one** that you inherited from your **father** present in the other member of the pair.

On the average, a single human chromosome contains an amount of DNA that, if fully stretched out, would be about 5 centimeters long! The total linear length of the DNA present in a single adult human's 37 trillion cells has been calculated to be the equivalent of nearly 70 trips from the earth to the sun and back. Imagine that — if you can! Not surprisingly, the DNA in each chromosome is extremely densely packed.

Sperm in males and ova (eggs) in females, called **germ cells**, contain half as many chromosomes — 23 **unpaired** chromosomes, with just one copy of every gene. When an ovum (egg) is fertilized by a sperm cell, the fertilized

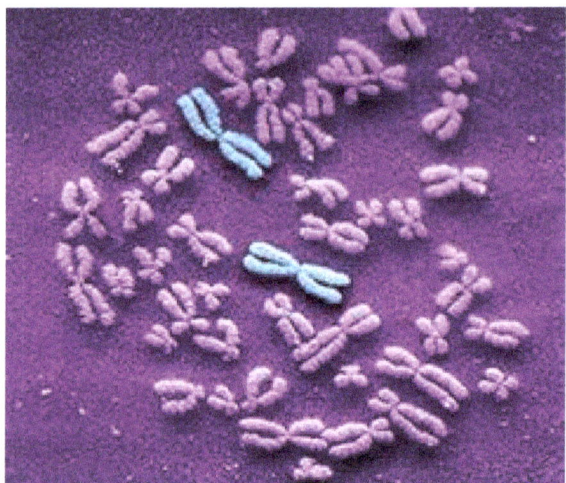

Fig. 5-1. Chromosomes can be isolated from the nuclei of cells, and when examined under a microscope can be counted, paired, and numbered. The two blue chromosomes are a pair.

ovum becomes a cell that is once again endowed with 46 chromosomes. As the fetus matures, germ cells with half the original amount of genetic information reappear.

You might wonder why germ cells (sperm and eggs) go through a process of halving their chromosome content, which they do during an intricate process called **meiosis**. If, like all other cells, our germ cells contained 46 chromosomes, every time a male sperm cell fertilized a female egg the number of chromosomes would double, and after just a few generations an early embryo would possess cells containing a gigantic number of chromosomes, an event that would lead to a gigantic genetic catastrophe! Mother Nature in her wisdom **invented the process of meiosis to prevent such a catastrophe**.

In addition to a reduction in chromosome number, **meiosis generates new combinations of DNA** in each of four daughter cells that are generated during this process. These new combinations result from exchanges of DNA between paired chromosomes, one from each parent. As a consequence of these exchanges the germ cells (male sperm and female ova) generated during meiosis acquire an astonishing range of genetic variation, which is why, though genetically similar to their parents, **all offspring are genetically distinct from either parent**.

* * *

When a fertilized egg grows to form an embryo, cells divide by a process called **mitosis**, which is distinct from meiosis in the sense that the genetic material is not rearranged as transpires during meiosis. During mitosis

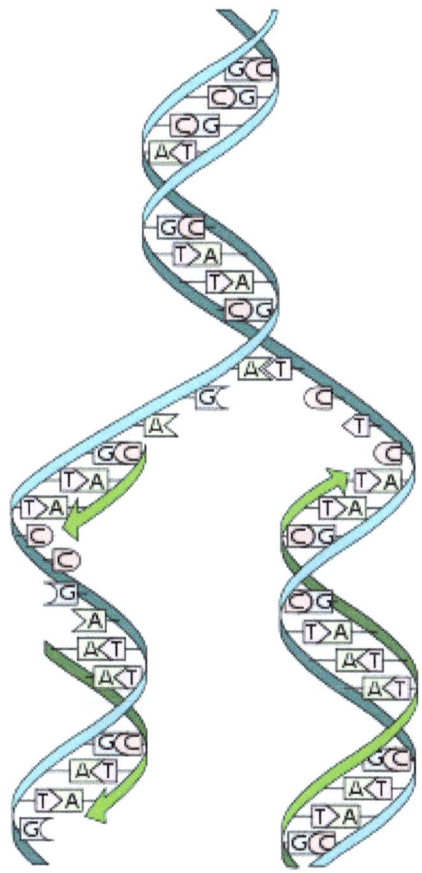

Fig. 5-2. During DNA replication double stranded DNA (at the top of the figure) is unzipped and unwound and each separated strand (turquoise color) acts as a template for replicating a new partner strand (green color). Bases are matched such that the base **A** in one DNA strand is always opposite the base **T** in the other DNA strand, and the base **G** is always opposite the base **C**.

a single cell divides, generating two identical cells, eventually leading to a fully formed embryo. The fact that an embryo ultimately consists of multiple cell types with different functions (brain cells, heart cells, liver cells, muscle cells, etc.) is a consequence of entirely different biological events that are unrelated to genes. Mitosis does not involve any changes in the genome. It transpires when cells in tissues and organs grow by division, thereby generating two identical daughter cells; for example during the growth of a fully formed embryo to an infant before it is born, or in adults during the healing of wounds. During mitosis, new DNA molecules are synthesized from precursors of the four bases A, T, G and C that are incorporated into new DNA strands by the action of an enzyme called **DNA polymerase** that facilitates the accurate incorporation of **A** opposite **T** and **G** opposite **C**, a process called **DNA replication** (Fig. 5-2).

As indicated in the figure, the end result of DNA replication is a new double-stranded DNA molecule with the identical base composition and order as the original DNA molecule. Remarkably, well before the enzyme DNA polymerase was discovered, the concluding sentence in Watson and Crick's famous article on the structure of DNA published in the journal *Nature* on April 25, 1953, prophetically stated: "It has not escaped our notice that the specific pairing between **A** and **T** and **G** and **C** — immediately suggests a possible copying mechanism for the genetic material."

6 The Genetic Code and Protein Synthesis

As already mentioned, the frequency and order of the bases A, T, C and G in the DNA of our genes embodies a code, referred to as the **genetic code; a code that instructs the synthesis (manufacture) of the multiple proteins in the cells of living organisms**. Proteins are made up of multiple small chemical units called **amino acids** of which there are 20. Each protein consists of different amino acids linked in different orders in chains called **polypeptides**, in much the same way that DNA consists of chains of four bases in different orders and lengths. Proteins may be comprised of a single polypeptide or multiple polypeptide chains. Once a polypeptide is completed it folds in a specific manner that generates a fully formed and functional protein **(Fig. 6-1)**.

The exact number of proteins in our cells is not known, but it is estimated that the human body has the ability to generate about **20–25 thousand different proteins**, encoded by 20,000–25,000 of our genes. Researchers have confirmed the existence of 19,599 protein-coding genes in the human genome and have identified another 2,188 DNA segments that are predicted to be protein-coding genes.

The unraveling of the genetic code and our current understanding of how polypeptide chains are manufactured is one of the most spectacular scientific triumphs in the history of biology, and represents one of the most elegant of Nature's many transactions! Several outstanding scientists contributed to deciphering the genetic code and comprehending the mechanism of protein synthesis. Following on from his work on DNA structure with Jim Watson, Francis Crick, together with **Sydney Brenner**, a brilliant young South African molecular biologist who relocated to

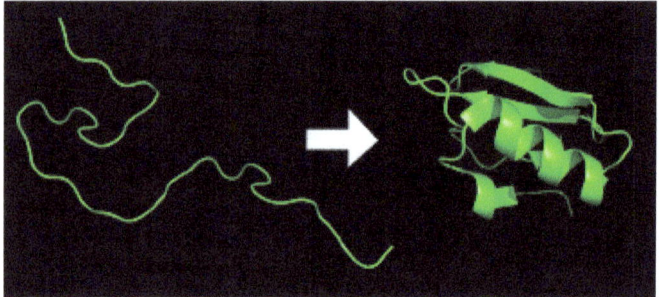

Fig. 6-1. Folding of a polypeptide chain to generate a functional protein.

Fig. 6-2. Sydney Brenner.

England to work with the leading figures in molecular biology, made seminal contributions to deciphering the genetic code and understanding how proteins are synthesized.

Brenner was born in 1927, in a small South African town called Germiston. His academic brilliance was noted at a very early age. He completed the first three years of primary school in one year and graduated from high school at the age of 14. Brenner entered medical school at the University of the Witwatersrand, in Johannesburg, South Africa and immediately became interested in biological research — to the virtual exclusion of medical subjects. He would be too young to legally qualify for the practice of medicine at the conclusion of his degree, so Brenner was allowed to complete a Bachelor of Science degree in Anatomy and Physiology. After graduating from medical school he completed a Ph.D. at Oxford University, returning to South Africa for a 2-year period stipulated by the scholarship he was awarded.

Like Sydney Brenner, I was born and raised in South Africa and attended the same medical school, where he lectured to my class in physiology during his mandatory 2-year return to the country. Aside from being an outstanding teacher, his lectures were typically laced with his inimitable sense of humor. In 2010 I published a biography of Brenner and in the course of that undertaking I got to know the man well and became a dedicated admirer.

One of many apocryphal stories that typify Brenner's humor involved an occasion when he was lecturing to a group of students in England. Smoking cigarettes in classrooms was permitted in those bygone days, and in the midst of lecturing Brenner, then a heavy smoker, popped a cigarette between his lips — wrong way round! The students of the course looked incredulously at one another with the delightful anticipation of his lighting the filter end of the cigarette, particularly after he also retrieved a lighter from one of his pockets. But continuing his lecture without breaking stride Brenner flicked on the lighter and nonchalantly turned the cigarette right way round before lighting it! At the time of this writing Brenner, aged 90, is plagued with frequent respiratory problems, presumably a consequence of his early habit of cigarette smoking. He resides in Singapore, where he spent considerable time and energy establishing molecular biology in that country.

In addition to his efforts in Singapore, Brenner helped establish the Okinawa Institute of Science and Technology (OIST) in Japan, a contribution for which he was awarded the *Order of the Rising Sun*, 1st Class, Grand Cordon by the Government of Japan. The *Order of the Rising Sun* is the highest award any scientist, Japanese or non-Japanese, can receive from the country.

* * *

Let's now proceed to examine the astonishing story of how the genetic code was deciphered or "cracked" (a descriptor often used in casual language), and how proteins are manufactured. As you now know, DNA resides in chromosomes in the nucleus of a cell and is manufactured there. But proteins were long known to be manufactured in the cytoplasm. As already mentioned, cytoplasm is a solution (mainly composed of water, salts, and proteins) that occupies the space between the nucleus and the cell membrane that fills each cell and is enclosed by the cell membrane **(Fig. 3-2)**. Since proteins were known to be made in the cytoplasm, whereas DNA is made in the nuclei of cells, Francis Crick, Sydney Brenner and several other investigators came to the conclusion that there must be a "messenger" molecule that transmits genetic information from the cell nucleus to the cytoplasm.

The notion evolved that the messenger molecule was made of a compound called **ribonucleic acid (RNA)**, which as noted earlier is chemically very similar to DNA except that **the base thymine (T) in DNA is replaced with one called uracil (U) in RNA**. It is not understood why this difference between DNA and RNA evolved. However, later in the book I'll mention a cogent suggestion that has been offered for this distinction.

Scientists had found notable amounts of RNA in a structure in the cytoplasm of cells called the **ribosome**, the known site of protein synthesis, and had tacitly assumed that ribosomal RNA was the postulated messenger. However, many previous experiments were not consistent with this idea. If ribosomal RNA was not the postulated messenger molecule, what was this elusive element? This question was resolved during a historic meeting at King's College, Cambridge on Good Friday, 1960, between a notable French molecular biologist named François Jacob, Sydney Brenner, Francis Crick and a handful of other scientists.

Some years earlier scientists working with a bacteriophage (that, as mentioned earlier, is a bacterial virus that infects bacteria) had discovered small amounts of RNA in bacteria infected with bacteriophage. This finding and its significance had remained unexplained. But, during the meeting at King's College, Brenner had the prophetic insight that this form of RNA is the elusive messenger because it copied the DNA composition of the bacteriophage, not the bacteria. This type of RNA, dubbed by Brenner as **messenger RNA (mRNA)**, was found in small amounts and had previously eluded detection because it was only required for short periods of time during protein synthesis. A series of brilliant experiments executed by Brenner, Crick and their colleagues proved Brenner's suggestion to be correct! This advance took the puzzle of how proteins are made a giant step forward.

Messenger RNA is made in cells by an enzyme called **RNA polymerase**, which works in much the same way that DNA polymerase does, always matching **C** in DNA with **G** in RNA, **G** in DNA with **C** in RNA, **T** in DNA with **A** in RNA and **A** in DNA with U in RNA. The process of synthesizing messenger (and other types of RNA) by RNA polymerase is called **transcription**, and the process of generating amino acid chains (polypeptides) during the synthesis of proteins is called **translation**.

With the basic concepts of protein synthesis in place, what remained to be explained was how information is **transcribed** from the bases **A, T, C and G** in DNA to the bases **A, U, C** and **G** in messenger RNA, which in turn is **translated** to protein. Brenner and his colleagues became persuaded that a fixed number of bases must encode the incorporation of a particular amino acid into a growing polypeptide chain. This concept became referred to as the "genetic code."

Prior to any experiments, a **triplet** genetic code (consisting of 3 consecutive bases in DNA) was anticipated based on simple arithmetic considerations. It was reasoned that the use of a **single** base code in which just A, T, C or G encoded a particular amino acid could obviously only encode four amino acids. Likewise, the use of two bases (AA, AT, AC, AG, etc.) could only encode 16 amino acids. But a 3-letter (triplet) code could theoretically specify 64 amino acids; more than enough to encode for the known 20 amino acids in Nature. Theorizing was one thing. But experimental proof that the genetic code consists of three bases (called a codon) was another thing altogether! A brilliant experiment executed by Crick, Brenner and their colleagues proved that the code embodied in a gene indeed consists of **three bases; a triplet genetic code. Figure 6-2** shows how Crick and Brenner experimentally established that the genetic code is indeed read in codons **consisting of three bases** instead of some other number such as four or five bases. Brenner and Crick carried out experiments of this type in which **X** was a chemical called **proflavine** that intercalates between any two nucleotides in DNA.

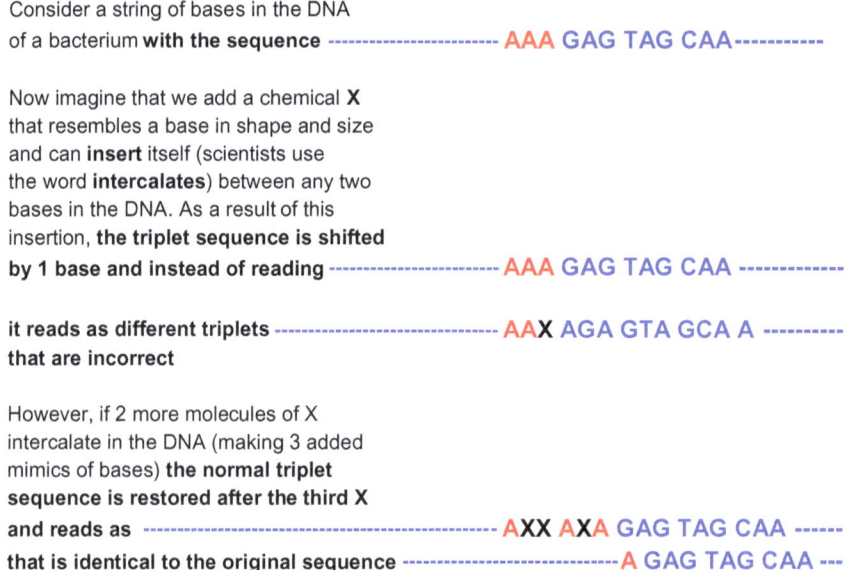

Consider a string of bases in the DNA of a bacterium **with the sequence** -------------------- AAA GAG TAG CAA-----------

Now imagine that we add a chemical **X** that resembles a base in shape and size and can **insert** itself (scientists use the word **intercalates**) between any two bases in the DNA. As a result of this insertion, **the triplet sequence is shifted by 1 base and instead of reading** -------------------- AAA GAG TAG CAA -------------

it reads as different triplets ------------------------------ AA**X** AGA GTA GCA A ----------
that are incorrect

However, if 2 more molecules of X intercalate in the DNA (making 3 added mimics of bases) **the normal triplet sequence is restored after the third X and reads as** -- A**XX** A**XA** GAG TAG CAA ------
that is identical to the original sequence ------------------------------A GAG TAG CAA ---

Fig. 6-3. Proving that the genetic code is a triplet code.

This experiment proved that the genetic code is read in triplets. If the genetic code were read by four adjacent nucleotides, then the events shown above would have required adding 4 molecules of **X**.

In his autobiography, *My Life in Science*, published in 2001, Brenner wrote:

> **It did not take long to realize that by shifting the phase of the code, essentially at will by using intercalating agents — we should be able to determine the size of a codon. If a gene was mutated because the codon reading frame had been shifted by 1 proflavine molecule, the total number of proflavine additions to get back into frame, into a sense register, should equal the number of nucleotides in a codon.**

If you had difficulty understanding with this experiment you are in good company! Brenner later wrote: "This concept of a phase shift — was so foreign to people in genetics that we had endless problems trying to explain this work." I like to refer to this brilliant experiment as "the 3 wrongs make a right" experiment!

Brenner and Crick called three sequential bases in a DNA strand **codons**. In addition to codons that specify a particular amino acid to be added to a growing polypeptide (protein) chain they discovered that certain codons signal the beginning and end of a protein chain.

In the opinion of many molecular biologists and geneticists, Sydney Brenner's contributions in the mid-1960s to understanding how the genetic code operates unquestionably deserved a Nobel Prize. But he had to wait until 2002 to earn that accolade, for another study that had nothing to with the genetic code. Perhaps the Nobel Committees that annually decide on the prizes were subjected to pressure to avoid the embarrassment of denying one of the great molecular biologists of the 20th century the award before he died? But this begs the question: why was Brenner not awarded the prize in the early 1960s to begin with? My own conjecture is that Brenner was (and still is) a flamboyant personality who does not suffer fools gladly and who can and not infrequently does offend without apology. The Nobel committee members on the other hand are likely staid Swedes. Therein lies a recipe for punishment!

* * *

The incorporation of amino acids into polypeptide chains to generate proteins does not transpire directly from reading the genetic code in mRNA. Messenger RNA recognizes a matching triplet nucleotide sequence in a **second type of RNA called transfer RNA (tRNA)** that is attached to the particular amino acid that will become part of the protein. And the new protein is assembled on yet **a third type of RNA**, the **ribosomal RNA** mentioned earlier.

I'll not trouble you with the details of these other types of RNA. All you really need to understand about the genetic code is that triplets of bases in DNA are matched to corresponding triplet bases in messenger (mRNA), which in turn recognize matching triplet sequences in another type of RNA called transfer RNA that is attached to a particular amino acid **(Fig. 6-4)**.

Fig. 6-4.

The incorporation of a particular amino acid into a growing polypeptide chain transpires on structures in cells called **ribosomes** that also contain RNA, which, as mentioned above, was initially thought to be the elusive messenger RNA. In essence, the genetic code relies on the inviolate rule that a triplet sequence in DNA, for example **CGA**, always recognizes the complementary sequence **GCU** in messenger RNA, which in turn recognizes **CGA** again in a second type of RNA (tRNA) that is hooked to a particular amino acid. Hence, it takes two sets of nucleotides **(GCU in mRNA and CGA in tRNA)** to restore the original coding triplet **CGA** in DNA **(Fig. 6-4)**.

You may ask why the **CGA** sequence in DNA does not *directly* recognize the sequence **CGA** in tRNA attached to an amino acid, skipping **GCU** in messenger RNA? As I related earlier, Mother Nature in her infinite wisdom placed DNA in the nucleus of cells and protein synthesis in the cytoplasm of cells. mRNA is required in the cytoplasm. Francis Crick referred to the series of events DNA→RNA→ protein as **"the central dogma"** of biological information. When an entire polypeptide chain of amino acids is fully assembled it folds into a unique three-dimensional structure that specifies that particular protein **(Fig. 6-5)**.

* * *

Once it was firmly established that the genetic code is a triplet nucleotide code, the next looming question was **which triplet codons specify**

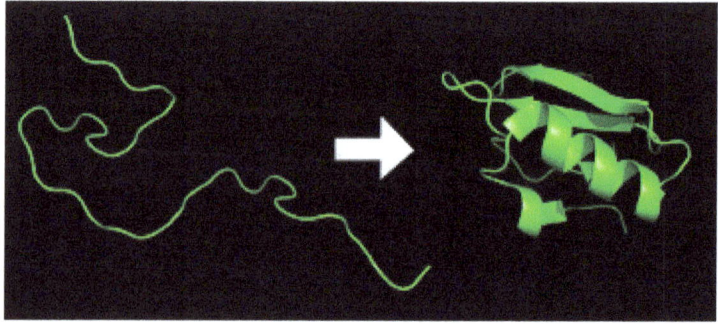

Fig. 6-5. Once a polypeptide chain (left) is fully completed it folds into a functional three-dimensional structure that is unique to that protein (right).

which amino acids? The first experimental evidence that answered this question was provided by Marshall Nirenberg.

Nirenberg was born in New York City, the son of Minerva Bykowsky and Harry Edward Nirenberg, a shirtmaker. He developed rheumatic fever as a boy, prompting the family to move to Orlando, Florida to take advantage of the subtropical climate. Nirenberg developed an early interest in biology and in 1948 received his B.S. degree, and in 1952, a master's degree in zoology from the University of Florida at Gainesville. His dissertation for the Master's thesis was an ecological and taxonomic study of caddis flies (*Trichoptera*). He received his Ph.D. in biochemistry from the University of Michigan, Ann Arbor in 1957.

Nirenberg began his postdoctoral work at the National Institutes of Health (NIH) in 1957 as a fellow of the American Cancer Society, in what was then called the National Institute of Arthritis and Metabolic Diseases. In 1959, he became a research biochemist at the NIH and began to study the steps that relate DNA, RNA and protein. In 1962 his groundbreaking experiments advanced him to the chiefship of the Section of Biochemical Genetics at the National Heart Institute (now the National Heart, Lung, and Blood Institute), where he, together with his postdoctoral fellow Heinrich Mattaei, made liquid extracts of bacterial cells (recall that bacterial extracts are made by crushing billions of intact cells and are used as the starting point to isolate a cellular component of interest or to support a biochemical reaction) and added a short piece of RNA with only **one kind of base, uracil (UUUUUUU)**, appropriately called **poly(U)**. He and his colleagues demonstrated that poly(U) RNA triggered the synthesis of a polypeptide chain comprised of a single amino acid called **phenylalanine**. By carrying out experiments of this kind using RNA with different combinations of bases, Nirenberg was able to decipher which triplet codons in DNA carried a code for which amino acids to be incorporated into a polypeptide chain.

* * *

One of the mainstays of a scientist's professional life is attending and participating in scientific conferences in various parts of the world. At such venues, scientists present the results of their work and hear presentations by others on topics of interest to them. In 1961, *The Fifth International*

Fig. 6-6. Marshall Nirenberg.

Conference of Biochemistry was held in Moscow, Russia. The conference drew over 5,000 scientists from around the world, including Nirenberg, who was looking forward to presenting the results of his exciting experiments revealing which codons instructed the appearance of which amino acids during protein synthesis. Given the large number of scientists presenting their research work at the 1961 meeting in Moscow, multiple presentations were scheduled simultaneously in different conference rooms, an unavoidable feature of large scientific meetings. Attendees at such meetings thus have the arduous task of deciding which talks they most want to hear, depending on their sphere of interest. Additionally, while talks at *The Fifth International Conference of Biochemistry* considered to be of major importance were selected for presentations of 20–30 minutes, the remainder were abbreviated to a mere 10 minutes. Such was the fate of Nirenberg's talk, which was attended by just a few dozen people.

In his book *Life's Greatest Secrets*, author Matthew Cobb related that Matthew Meselson, a recognized Harvard University scientist, attended Nirenberg's talk and was so excited by his presentation that among the many hundreds of scientists milling around the conference facility, he managed to track down Francis Crick and informed him of the exciting news. Crick was to serve as the chairperson of one of the scientific sessions scheduled was for the following day that featured 20–30 minutes presentations and after hearing about the content of Nirenberg's 10 minutes talk from Meselson he invited Nirenberg to give his presentation again the following day, this time in a 20-minute talk.

Author Matthew Cobb quotes Nirenberg as later stating: "The second time I gave the paper it was to a very large audience. The reception was really remarkable, fantastic. I remember Matt Meselson, who was sitting right up front. I didn't know him at the time, but he was so overjoyed about hearing this stuff that he impulsively jumped up, grabbed my hand, and actually hugged me — I could have been part of a rock band!"

The genetic code was now fully revealed to an extent that it was understood which triplet codons in DNA stipulated which amino acids become part of a growing protein chain.

7 Splicing Genes

You will presumably be surprised to learn that your DNA does not consist of a long uninterrupted sequence of bases that exclusively code for the synthesis of chains of amino acids that become proteins. Non-coding regions of DNA called **introns** interrupt the coding regions, which are called **exons**. During transcription to generate messenger RNA (mRNA) the entire gene is first transcribed to **pre-mRNA**, which includes both exons and introns. During a process called **RNA splicing**, introns are removed from the genome and the exons are joined to form contiguous coding sequences. The mature mRNA is then available for translation.

The intron–exon architecture of many genes raises the intriguing question of whether this unique biological organization serves any function, or whether it's simply a result of the spread of functionless introns in our genomes during evolution. Evidence in favor of functionality is considered likely for several reasons. For one thing, the presence of introns in genomes is thought to impose substantial burdens on cells. The splicing (excision) of introns requires a complex structure called the **spliceosome**, which is among the largest molecular complexes in the cell, comprising multiple types of RNA and more than 150 proteins. Additionally, the transcription of introns (converting the intron sequences in DNA to RNA) is costly in terms of energy spent. Finally, the accurate recognition of splicing junctions by the spliceosome is directed by a large number of regulatory elements. This biochemical complexity renders an organism vulnerable to mutations that otherwise would not have a noticeable effect. In fact, it has been suggested that more than 50% of human genetic disorders are caused by disruption of the normal pattern of slicing out introns.

Why did Nature invent introns in your DNA? For one thing, it has been shown that the removal of introns decreases the amount of messenger RNA in cells and hence the amount of manufactured protein. In some cases intron-bearing DNA made in the laboratory expresses up to 400 times more mRNA than intronless DNA constructs. Some introns are so efficient in boosting expression levels that they are routinely included in DNA constructs made in the laboratory in order to guarantee high levels of expression of mRNA. Additionally, **alternative splicing** (splicing at different places in a gene) is a regulated process that allows **a single gene to code for multiple different proteins**; another example of Nature's exquisite biological devises.

As mentioned several times, only about 2% of our DNA encodes the instructions for making proteins. What is the other 98% for? Believe it or not, in 2017 we have to admit that we don't know! There was a time when scientists thought that 98% of the genome does nothing and it was derisively referred to as "junk" DNA! But it's distinctly unsatisfying to accept that 98% of our DNA has no function, otherwise it would have long since disappeared. Scientists are well aware of how useful it would be to understand what the other 98% of our genome is for. But as yet no practical way of achieving this goal has been generated, presumably in large part because we don't know what function(s) to search for!

A comprehensive appreciation of the millions of years that have passed since the beginning of evolution is difficult to acquire. But I hope that your comprehension of the multitude of biochemical events and cellular structures required to make proteins has given you an appreciation of the time it took to arrive at the elegant strategy that ultimately arose on Earth. Perhaps there were eons in the dim and distant past when proteins were made by some other mechanism(s). But, as new and more efficient ways that afforded advantages over previously existing strategies evolved, earlier ones disappeared. Consider the evolution of travel, first by walking and running, then in primitive carts without wheels, followed by carts with wheels, rowing boats, bicycles, vehicles with wheels, motorized vehicles, electrically powered ships and ultimately airplanes. That technical evolution took a long time. What's next in store for humanity? Certainly driverless cars, and perhaps even cars that fly! Ultimately, possibly even instant transformation of matter from place A to place B! But such technical evolutions are mere grains in the sands of the time required for biological evolution as we know it today.

* * *

When a gene is actively directing the synthesis of a particular mRNA we refer to the gene as being **expressed**. But all the genes in your genome do not simultaneously express information that leads to the formation of proteins. Each cell type in your body has a different set of expressed genes and different patterns of expression cause different cell types to make different types of proteins. An example that I discovered on the web reminds us that one of the jobs of the liver is to remove toxic substances like alcohol from the bloodstream. To accomplish this, liver cells express (**turn on**) genes that encode components of an enzyme that changes alcohol to a non-toxic molecule. But cells in your brain don't remove toxins from the body, so the brain keeps these genes repressed (**turned off**). Cells also possess complex mechanisms for altering the expression of genes. As you have learned, the expression of genes involves multiple steps, and almost all of them can be regulated. Gene regulation is a multifaceted topic that is not considered in this book.

8 Damaged and Incorrect DNA can (Sometimes) be Repaired

You might reasonably believe that the DNA in your cells is well protected in its cocoon of chromosomes, snuggly buried in the nuclei of your cells. Regrettably, such is not the case. DNA is a highly reactive biological compound that sustains various types of damage and alterations by interacting with chemicals that are generated during normal metabolic processes or by interaction with man-made chemicals that enter our bodies. Some of the chemicals that arise during metabolic processes are referred to as **R**eactive **O**xygen **S**pecies **(ROS)**. ROS are the products of normal oxygen-consuming metabolic process in the body that are highly reactive with DNA. You may be shocked to learn that **in humans, naturally occurring oxidative DNA damage has been calculated to occur at least 10,000 times per cell per day**. That's a lot of DNA damage!

DNA damage can also arise from spontaneous breaks in one or both DNA strands, or as the result of the incorporation of incorrect nucleotides by the enzyme DNA polymerase during the process of DNA replication (the copying of existing DNA chains to generate new DNA molecules, a topic discussed in a later chapter). These multiple sources and modes of damage to our DNA can interfere with DNA replication and/or transcription (another topic discussed later), events that can lead to cell death. The good news is that over the eons of biological evolution, biochemical mechanisms evolved that promote the **repair of damaged DNA**, including the repair of incorrect nucleotide pairs generated by faulty DNA replication.

The first known (and possibly the oldest) DNA repair mechanism was accidently discovered in the mid-1940s by a scientist named Albert Kelner. Kelner was born in 1912. In his early teens he was afflicted with tuberculosis of the bone, a disease that left him with a permanent and pronounced limp. His left shoulder was also affected and interfered with his ability to play the violin, an instrument that he once performed on with considerable skill. Kelner's wife once informed me that had her husband not developed an early interest in biology he almost certainly would have become a musician. "We might have starved as a family" she said, "but that's beside the point. He was a very fine musician."

Kelner's discovery of DNA repair transpired during a fateful encounter with antibiotics, compounds produced by bacteria and fungi that are capable of killing or inhibiting competing microbial species. This phenomenon has long been known and may explain why even the ancient Egyptians had the practice of applying a poultice of moldy bread to infected wounds. But it was not until 1928 that penicillin, the first known antibiotic, was discovered by Alexander Fleming, Professor of Bacteriology at St. Mary's Hospital in London, for which he received a Nobel Prize in 1945.

Returning from a vacation in September 1928 Fleming began to sort through Petri dishes containing colonies of *Staphylococcus*, a bacterium that can cause serious infections. While so engaged, he noticed something unusual on one dish. It was dotted with bacterial colonies just like the other dishes, save for one area where a blob of mold was growing. The zone immediately around the mold was clear, as if the mold had secreted something that inhibited bacterial growth. Fleming was intrigued by this observation and found that his "mold juice" was capable of killing a wide range of harmful bacteria. He had discovered penicillin!

During World War II, in addition to the mass production of penicillin in the United States, an intensive search was launched for new antibiotics. Between 1943 and 1946 Albert Kelner was at the University of Pennsylvania, where he carried out mass screenings of bacteria in the hope of discovering new antibiotics. In 1946 he secured a position at the Cold Spring Harbor Laboratory, then under the direction of Milislav Demerec, a Croation-born scientist who was in search of new antibiotics. Demerec wanted to explore the interesting notion that microorganisms not known to excrete antibiotics might be mutated to do so. When Kelner joined Demerec's laboratory he was assigned this task.

In order to generate bacterial mutants, one or more of which would hopefully produce new antibiotics, Kelner exposed bacteria to ultraviolet

(UV) light, a frequently used method of generating mutations in bacteria, since UV light is highly reactive with and damaging to DNA. To calibrate his experimental system, Kelner exposed *E. coli* bacteria to various doses of UV light to establish the amount of exposure that yielded the optimal level of mutated bacteria with the least amount of killing.

To his considerable frustration Kelner obtained results that were annoyingly irreproducible. By dogged persistence he eventually identified fluorescent lights in the laboratory as the source of his variable results and established that cells that had been subjected to UV radiation experienced significant recovery if they were subsequently exposed to visible light. Kelner soon realized that he had discovered a biological process called **DNA repair**. Excited at having stumbled onto this discovery, Kelner, with Demerec's permission, aborted his search for bacteria that might produce new antibiotics and focused on the phenomenon of DNA repair.

By extraordinary coincidence, at the very time that Kelner was trying to make sense of his baffling observations when he exposed bacteria to UV light, **Renato Dulbecco**, an Italian postdoctoral fellow at the University of Indiana was carrying out experiments on the effect of UV radiation on bacteriophage. Like Kelner, Delbecco was exasperated to observe that when he stored stacks of agar plates (on which bacteria grow as visible colonies), the number of phage survivors varied enormously. Eventually he realized that the agar plates on top of the stacks (and consequently exposed to visible light) had the greatest number of phage and he too comprehended that he had discovered a light-dependent DNA repair mechanism that he called **photoreactivation**. In due course Kelner and Dulbecco corresponded and agreed to share credit for the discovery of photoreactivation — and of DNA repair.

The focus on enzymatic photoreactivation soon shifted to understanding what transpires during this mode of DNA repair. To comprehend enzymatic photoreactivation we need to revisit the bases thymine (T) and cytosine (C). These two bases belong to a chemical family called **pyrimidines**. It was well known that when DNA is exposed to UV light, two adjacent pyrimidines in a strand of DNA can be chemically joined to generate a type of DNA damage called **pyrimidine dimers (T-T) (T-C) and (C-C)**. Pyrimidine dimers alter the three-dimensional structure of DNA, resulting in arrested DNA replication by DNA polymerase and reduced levels of transcription of genes by RNA polymerase. However, when DNA

Fig. 8-1. Renato Dulbecco.

containing pyrimidine dimers is exposed to **visible** light, an enzyme called **photoreactivating enzyme** converts the dimers to normal unlinked TT, TC or CC bases. Regrettably, photoreactivating enzyme is not present in mammals. If our cells were endowed with this enzyme we might enjoy a much reduced risk of skin cancer after lying out in the sun, day after day!

In subsequent years, a number of other DNA repair mechanisms were discovered, including one called **excision repair**, a major DNA repair process that recognizes a wide range of damaged bases in DNA and excises (cuts out) the damaged segment of the DNA strand **(Fig. 8-2)**. This leads to the generation of gaps in the affected DNA strand that are filled in by DNA polymerase, that uses the opposite normal DNA strand as a template, just as transpires during the replication of DNA **(Fig. 8-2)**. The process of excision repair is not confined to the removal of thymine dimers. The excision repair machinery recognizes and repairs multiple different types of base damage in DNA. Scientists therefore surmise that the excision repair machinery recognizes a distortion of the structure of DNA that is common to multiple types of base damage. The nature of this putative distortion remains to be identified.

You will hopefully recall that whereas DNA contains the base **cytosine (C)**, RNA contains the base **uracil (U)** instead. I mentioned earlier that scientists don't know why this is the case. A reasonable explanation for this oddity is the following.

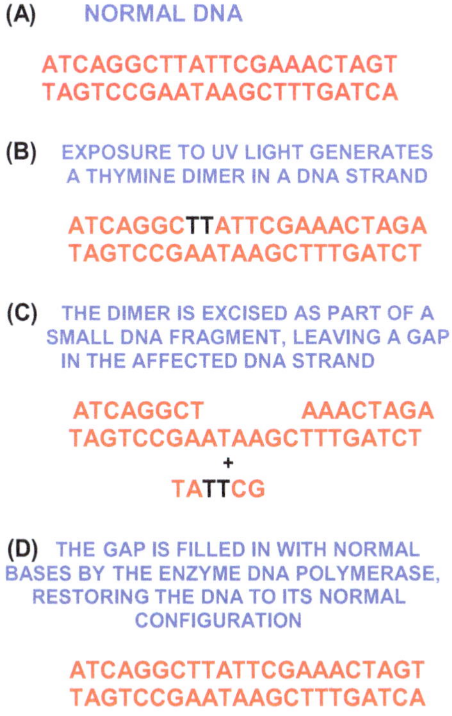

(A)   NORMAL DNA

ATCAGGCTTATTCGAAACTAGT
TAGTCCGAATAAGCTTTGATCA

(B)   EXPOSURE TO UV LIGHT GENERATES
A THYMINE DIMER IN A DNA STRAND

ATCAGGC**TT**ATTCGAAACTAGA
TAGTCCGAATAAGCTTTGATCT

(C)   THE DIMER IS EXCISED AS PART OF A
SMALL DNA FRAGMENT, LEAVING A GAP
IN THE AFFECTED DNA STRAND

ATCAGGCT   AAACTAGA
TAGTCCGAATAAGCTTTGATCT
+
TA**TT**CG

(D) THE GAP IS FILLED IN WITH NORMAL
BASES BY THE ENZYME DNA POLYMERASE,
RESTORING THE DNA TO ITS NORMAL
CONFIGURATION

ATCAGGCTTATTCGAAACTAGT
TAGTCCGAATAAGCTTTGATCA

Fig. 8-2.   Excision repair of DNA.

The chemical structures of the bases **cytosine (C)** and **uracil (U)** are very similar. Consequently, **cytosine (C)** in DNA can sometimes spontaneously change to **uracil (U)**. This change is more frequent when cells are exposed to certain chemicals. Tomas Lindahl, a Swedish-born chemist, discovered a **DNA repair mechanism** during which **(U)** in DNA (but not in RNA) is selectively removed by a repair enzyme and is replaced by **cytosine (C)**. However, if DNA normally contained **(U)** instead of **(C)**, the DNA repair strategy that removes **(U)** from DNA would remove **U** from undamaged DNA.

* * *

The last DNA repair mechanism I'll relate is one called **mismatch repair**. Think back to my informing you that during the synthesis of new DNA strands by the enzyme DNA polymerase an incorrect base can sometimes be incorporated in the new strand, generating **mismatched base pairs**. Mismatches are known to be repaired by a process during which the incorrect base is excised from the DNA and is replaced by the correct one. But operating under the reasonable assumption that the repair process recognizes

Fig. 8-3. Paul Modrich.

Fig. 8-4. Tomas Lindahl.

mispaired bases in DNA, a troubling question arose. Since both bases in the incorrect pair are chemically normal, **how does the repair process "know" which of the two DNA stands contains the incorrect base**?

Convinced that a mechanism existed for distinguishing the normal DNA strand from the newly synthesized one containing the mispaired base, this conundrum was solved by Matthew Meselson, a scientist at Harvard University who discovered that when DNA is copied by DNA polymerase to generate a new DNA strand, **the template DNA strand that is being copied is transiently chemically modified, while the newly synthesized daughter strand is not**. This nuance functions as a signal that "instructs" the mismatch repair process to remove the nucleotide in the unmodified strand, regardless of what that nucleotide is.

Paul Modrich at Duke University elucidated the details of the biochemical mechanism by which incorrect bases are removed during mismatch repair and replaced with correct nucleotides. In 2015, Modrich was awarded a Nobel Prize, together with Tomas Lindahl mentioned earlier.

9 Mutations can Cause Diseases

As you now hopefully comprehend, when synthesizing new DNA the enzyme DNA polymerase "knows" that the base A must always be placed opposite T, and C must be placed opposite G. As mentioned in the previous chapter, DNA polymerase (that like all enzymes is not perfect in its function) sometimes incorporates an incorrect nucleotide during DNA synthesis — thankfully very rarely. For example, instead of the base **A** being incorporated opposite **T** in a growing chain of DNA, the incorrect base **C** or **G** or even another **T** may be added to the growing chain. **Incorrect or unrepaired damaged bases in genes can generate mutations that may interfere with the normal function of affected genes**, sometimes leading to serious diseases — such as cancer.

Whether or not a mutation has a deleterious effect on an individual depends on how it affects the involved gene. If a mutation occurs in a region of a gene that is not essential for its function it may have no observable effect (so-called **silent mutations**). Or the mutated gene may slow down a biochemical reaction, causing only a minor disturbance in gene function.

Mutations in genes other than germ cells are called **acquired mutations** and may affect the normal function of an organ to varying degrees, but cannot be passed on to offspring. In contrast, mutations in germ cells (sperm and eggs) can be passed on to offspring and may lead to genetic diseases. Such mutations are called **hereditary mutations** and the diseases they may cause in offspring are typically incurable. However, encouraging progress is being made in this situation, generating a field of scientific endeavor called **gene therapy**.

It's instructional to understand that mutations in germ cells do not always have a deleterious outcome. In particular, the evolution of new species depends on non-deleterious mutations in germ cells. You might think that there are large genetic differences between humans and chimpanzees. But in fact the two species share about 99% of their DNA. It's believed that humans and chimpanzees shared a common ancestor around 10 or 11 million years ago. Hence, over a time span of 10–11 million years, only about 1% of our and chimpanzee's DNA changed. But that's a lot of change in genetic terms — which is why it's not difficult to pick out a single human among a crowd of chimps — or vice versa!

* * *

Hereditary mutations are at the seat of a number of important genetic diseases, including **cystic fibrosis (CF)**, **Huntington's disease** and a number of other genetic disorders. Those afflicted with CF, a relatively frequent disorder, inherit a defective gene on chromosome number 7 called *CFTR — Cystic Fibrosis Transmembrane conductance Regulator*). Cystic fibrosis is a genetically **recessive disorder**, meaning that both parents must pass on the defective gene for any of their children to have the disease. (Think back to Gregor Mendel, who noted the difference between dominant and recessive "factors".) If a child inherits only one copy of the faulty gene, he or she will be a **carrier**. Carriers don't manifest the disease but they can pass the mutation to their own children. Multiple genetic diseases have this feature.

The protein produced by the *CFTR* gene normally helps salt (sodium chloride) move in and out of cells. However, if the protein is dysfunctional due a mutation in the *CFTR* gene, the movement of sodium chloride across the cell membrane can be disturbed, resulting in the generation of abnormally thick sticky mucus on the outside of cells. Cells in the lungs are most seriously affected. The mucus clogs the airways in the lungs and increases the risk of infection by bacteria. The thick mucus also blocks passages in the pancreas, interfering with the access of pancreatic digestive enzymes to the intestines. Absent these enzymes food is not properly digested and people who suffer from CF are often starved of the nutrition required to grow normally. CF also affects sweat glands. The loss of too much salt through sweat can disrupt the delicate balance of minerals in the body.

The genome locations of many disease-related genes are not known. Hence, considerable efforts have been devoted to mapping disease-related genes in the genome. Mapping human genes can be achieved in two ways referred to as **physical mapping** and **genetic mapping**. Physical maps are based on an alignment of specific DNA sequences. Genetic mapping, which is less precise, involves the establishment of the relative position of a gene in the genome. In his recent book *DNA: The Story of the Genetic Revolution*, James Watson offers a simple analogy of genetic mapping that I'll borrow here.

"The technique, called **linkage analysis**, determines the position of a gene in relation to the known positions of particular genetic landmarks. The principle is simple: it would be difficult for you, given no other information, to find Springfield, Massachusetts on a map of the United States. But if I tell you that Springfield lies about halfway between New York and Boston — two landmarks labeled on the map — then your task is made very much easier. Linkage analysis aims to do this with genes: it establishes links between known genetic markers and unknown genes." Watson informs that "the dearth of known genetic markers in humans prevented its application to human diseases — until" two investigators, David Botstein and Ron Davis, realized that DNA sequences called restriction fragment length polymorphisms (RFLPs) constitute excellent genetic markers. RFLPs "occur when a DNA sequence cut by a particular enzyme in one individual has changed in another individual so that it can no longer be cut by that enzyme. —Millions of them are scattered throughout our genome."

"Botstein, Davis and several other collaborators eventually laid out a clear plan showing how RFLP markers could be used to produce a map of signposts up and down every human chromosome. Botstein and his colleagues calculated that 150 RFLPs spread uniformly across the entire genome would be enough to enable researchers to pinpoint mutated genes that cause disease. Collecting DNA samples from large affected families in which a disorder spanned several generations, they would track the inheritance pattern of RFLPs one after another; looking for ones that tracked the disease through the families, thereby signaling the approximate location of the mutated gene. Several years later, the geneticist Helen Donis-Keller published an article entitled *A Genetic Linkage Map of the Human Genome*. The map included 403 markers — spanning a good 95% of the human genome."

"In subsequent studies one of these genetic markers tracked with Huntington's disease, an inherited genetic disease that causes the progressive degeneration of nerves located inside the brain. The disease typically causes cognitive, psychiatric and movement disorders. In March 1993, a consortium of 59 authors announced that they had located the Huntington's gene and identified the mutation responsible for the disease. Nothing about the sequence of its 10,000 nucleotides suggested what its protein product might normally do in the brain or the body."

"Fortunately, genetic diseases are not common compared to the many disorders with no genetic component. Additionally a number of birth defects and genetic disorders can be tested for if a pregnant woman wishes to have this information, and screening tests can be performed prior to pregnancy. It is generally recommended that pregnant women who are members of families with a history of genetic disease consult their obstetrician about the availability of screening and diagnostic tests."

10 DNA Sequencing, Gene Cloning, Recombinant DNA Technology, DNA Fingerprinting and Gene Therapy

Recent decades have witnessed the emergence of a number of technologies that have facilitated deeper insights about genes and DNA.

DNA Sequencing

DNA sequencing is a technology that allows determination of the precise order of the four bases A, G, C and T in a DNA strand of any length. The technology is complex and will not be described here. As you might imagine, the advent of DNA sequencing greatly accelerated biological and medical research. As discussed in the next paragraph, knowledge of DNA sequences has become indispensable for basic biological research and for numerous applied fields. The rapid speed of sequencing attained with modern DNA sequencing technology has been instrumental in the sequencing of complete genomes of numerous types and species of life, including humans.

The most useful information that derives from DNA sequencing stems from information about the functions of particular DNA sequences (typically genes) by comparing them to the sequences of elements with known functions. A good example concerns genes that encode proteins required for the sense of smell. In vertebrates, proteins called olfactory receptors facilitate the sense of smell. In mammals, the genes that encode olfactory receptors constitute a large family of genes. Very few of the proteins encoded by these genes have been directly studied in laboratory experiments. But scientists know that these genes encode olfactory receptors because they share common DNA sequence components. As databases of DNA sequences continue to grow in volume, so will opportunities for establishing their functional significance by comparing them to matching sequences in genes with known functions.

Gene Cloning

In relatively recent times, technologies have been developed for isolating and studying individual genes, a procedure called **gene cloning**. Gene cloning is commonly used to isolate individual genes in order to investigate their characteristics and function. Additionally, one can introduce mutations at selected places in genes to determine how their function may be altered.

Cloning genes in organisms such as bacteria is relatively simple. Let's consider the phenomenon of defective DNA repair, that results in the inability of bacteria to grow when exposed to UV light because they are mutated in a gene required for DNA repair, a state that we call the **phenotype** of those bacteria. One can generate libraries of bacterial DNA in which individual bacteria contain different pieces of the bacterial DNA joined to a different DNA called plasmid DNA. This joining is achieved using enzymes that "cut and paste" DNA, and results in the generation of a plasmid library that carries different pieces of the bacterial DNA. Plasmids naturally exist in bacteria, in which they multiply. One infects the mutant bacteria that are unable to grow when exposed to UV light with the plasmid DNA and exposes these bacteria to UV light. Any bacteria containing plasmids that carry genes required for the repair of DNA damaged by exposure to UV light will survive the exposure to UV light. The DNA in these bacterial colonies can then be used to retrieve the genes of interest.

One can similarly clone human genes, using human cells that are defective in biological activities related to the genes for which one wants to clone. When I operated a laboratory at Stanford University, I studied DNA repair in human cells and used cells defective in DNA repair resulting from a disease called xeroderma pigmentosum (XP). People with mutant XP genes (of which there are several) are particularly sensitive to sunlight and frequently develop skin cancer. My colleagues and I were able to clone several genes, defects in which resulted in XP.

In addition to the utility of cloned genes for the study of gene function, cloned genes can be introduced into bacteria and the protein that the gene encodes can be harvested for commercial use. For example, let's assume that you own a company in which you want to manufacture and sell human insulin for the treatment of diabetes. Insulin is a protein made by cells in the pancreas and without it one suffers from diabetes. Just as I described the procedures one would use for cloning genes required for DNA repair, you could clone the human insulin gene (or perhaps acquire or purchase it) and insert it into a plasmid. When you infected bacteria with your **recombinant plasmid DNA** billions of copies of plasmid DNA that manufacture human insulin would be generated. If you executed these maneuvers on an industrial scale you would be able to harvest kilogram quantities of the plasmid-bearing bacteria. After crushing the bacteria you would have an extract from which you could purify large quantities of insulin.

Recombinant DNA Technology (Genetic Engineering)

Whereas gene cloning refers to the isolation of a gene of interest from the genome of an organism, **recombinant DNA technology**, often referred to as **genetic engineering**, involves the introduction of cloned genes into the genome of a different organism.

In 1972, Paul Berg at Stanford University created the first recombinant DNA molecules by combining DNA from the monkey virus SV40 with that of a bacteriophage that infects the bacterium *E. coli*. Berg contemplated introducing this recombinant DNA into a laboratory strain of *E. coli*. However, he did not complete this step because of concerns raised by scientists fearful of potential biohazards. Specifically, SV40 virus was known to cause cancer in mice. Additionally, *E. coli* (although not the strain used by Berg) was known to inhabit the human intestinal tract. Hence, some feared that bacteria containing SV40 DNA might escape into the environment and infect laboratory workers (and others), who might become cancer victims.

Subsequently, Herbert Boyer at the University of California in San Francisco and Stanley Cohen also at Stanford created the first transgenic organism by inserting antibiotic resistance genes into a plasmid of *E. coli*. And Rudolf Jaenisch at the Massachusetts Institute of Technology (MIT) created a transgenic mouse by introducing foreign DNA into its embryo, making it the world's first transgenic animal.

These achievements led to concerns in the scientific community about potential risks from genetic engineering. Here was a technology by which any piece of DNA could in theory be introduced into bacteria such as *E. coli*, which in turn could infect people. This raised apprehension in the scientific community because SV40 can transform monkey and human cell lines to a cancerous state.

These concerns prompted Berg to convene a conference chaired by Berg at Asilomar, California in 1975. A protracted period of debate that centered on the safety of the technology followed. Subsequently rules governing the generation, storage and handling of some of the products of recombinant DNA technology were implemented around the world. At the time of this writing no incidents in which the nightmare scenario of dangerous DNA running amok in and outside laboratories have transpired.

DNA Fingerprinting

DNA fingerprinting, also called DNA typing, DNA profiling, genetic fingerprinting, genotyping, or identity testing, is a method for identifying elements in the sequence of DNA **that are unique to every individual in the world**, hence the use of the term fingerprinting. The technique was developed in 1984 by a British geneticist named Alec Jeffreys (now Sir Alec Jeffreys) who, while working at the University of Leicester, observed that certain sequences of highly variable DNA (known as **minisatellites**) that do not contribute to gene function, are repeated within genes. Pursuing these studies further, Jeffreys recognized that every individual in the world has a **unique pattern of minisatellites**, the only exceptions being multiple individuals from a single fertilized egg, such as identical twins.

In a recent interview Jeffreys stated: "My life changed on Monday morning at 9.05 am, on September 10, 1984. What emerged was the world's first genetic fingerprint. In science it is unusual to have such a 'eureka' moment. We were finding extraordinarily variable patterns of DNA, including from my laboratory technician and her mother and father, as well as from non-human samples. My first reaction to the results was that this is too complicated, and then the penny dropped and I realized that we had genetic fingerprinting."

In 1985, Jeffreys and his team developed a variation of DNA fingerprinting for forensic use, which is based on the same principle. Its first application was in a murder case that is enthrallingly recounted in a book called *The Blooding*, by Joseph Wambaugh, published in 1989, a book that I recommend to readers interested in the first use of DNA fingerprinting in a criminal investigation.

On November 21, 1983, a 15-year-old girl named Lynda Mann left her home in the Leicester area of England to visit a friend's house. She did not return. The next morning she was found raped and strangled on a deserted footpath. Using forensic science techniques available at the time, police linked a semen sample taken from her body to a person with type A blood and an enzyme profile that matched only 10% of males. With no other leads or evidence, the case remained open.

On July 31, 1986, another 15-year-old girl, Dawn Ashworth, also from the Leicester area, took a shortcut walk home instead of her normal route. Two days later her body was found in a wooded area near a footpath. She had been beaten, savagely raped and strangled to death. The *modus operandi* matched that of the first attack and semen samples revealed the same blood type as that obtained from the examination of the forensic evidence obtained from Lynda Mann's body.

The prime suspect, a 17-year-old youth named Richard Buckland who suffered from learning difficulties, revealed knowledge of Ashworth's body and admitted the crime under questioning, but denied the first murder. Alec Jeffreys compared semen samples from both murder victims against a blood sample from Buckland and conclusively proved that both girls were killed by the same man, **but not by Buckland**! Buckland became the first person to have his innocence established by DNA fingerprinting. Jeffreys later stated: "I have no doubt whatsoever that he would have been found guilty had it not been for DNA evidence. That was a remarkable occurrence."

Given the availability of Alec Jeffreys's fingerprinting test the Leicestershire Constabulary undertook an investigation during which more than 5,500 local men were asked to donate blood or saliva samples — an extraordinary event! This massive screening took six months to complete, at the end of which no matches with the semen samples were found!

Enter Colin Pitchfork, a young married man with two sons who lived in the Leicester area. Pitchfork worked at a local bakery. He became particularly skilled as a sculptor of cake decorations and had hoped eventually to start his own cake decorating business. According to his supervisor he was "a good worker and time-keeper, but he was moody — and he couldn't leave women employees alone. He was always chatting them up." Pitchfork had once been convicted of indecent exposure and had been referred for therapy at a local hospital.

In August 1987, Ian Kelly, one of Pitchfork's colleagues at the bakery where they both worked, revealed to fellow workers in a Leicester pub that he had obtained 200 English pounds from Pitchfork for giving a blood sample while masquerading as Pitchfork. Pitchfork had told Kelly that he could not give blood under his own name because he had already given blood while pretending to be a friend who had wanted to avoid being harassed by police because of a youthful conviction for burglary. Police quickly uncovered this ruse by Pitchfork and arrested him for the murder of both women. During subsequent questioning, Pitchfork admitted exposing himself to more than 1,000 women, a compulsion that had begun in his early teens. He later progressed to sexual assault. He pleaded guilty to the two rape/murders and was sentenced to life imprisonment.

When asked how he felt when Pitchfork was finally convicted, Alec Jeffreys is quoted as saying: "I felt relief because he was a serial murderer

Fig. 10-1.   Alec Jeffreys.

and would kill again, and because if the search had failed, the public's perception of forensic DNA would have been shattered. Also, here was a serial killer in the region who knew what I was doing and where I worked and where my family lived. That felt very uncomfortable, so on a personal level it was a great relief when he was trapped."

Gene Therapy

As mentioned earlier, the day may well arrive when individual functional genes may be introduced into humans who suffer from diseases that are the result of defective genes, as a cure for genetic diseases — so-called **gene therapy**. In recent years, techniques for the transfer of genes from the test tube to target organs in people with defective genes have gained increasing acceptance in medical and scientific circles. Several ways of accomplishing gene therapy are under active investigation as I write. Besides replacing mutant (defective) genes with healthy ones, these strategies include shutting down the expression of mutant genes.

Gene therapy is now more than just wishful thinking. In 2016, a group of Italian scientists reported that they had cured a number of children suffering from a genetic deficiency in the immune system. They removed the children's bone marrow, added a gene to make the enzyme that their bodies lacked and replaced the bone marrow. In early 2017 the anticipated cost of this gene therapy was stratospheric! However, there is little doubt that in time, more precise and affordable strategies for altering genes that promote genetic diseases will come to fruition.

"One much talked about potential strategy called **CRISPR** (an acronym that I won't bother to translate) is a type of bacterial self-defense mechanism that evolved in bacteria to recognize and destroy viral invaders," Jim Watson wrote. "**CRISPR** piggybacks on an immune strategy that bacteria use to detect and chop up foreign DNA. The DNA-cutting enzyme Cas9 finds its gene target with the help of an RNA guide sequence that researchers can now engineer to home in on potentially any gene of interest. Based on this information, there are optimistic indications that the technology can be used to cut human DNA with exquisite precision, potentially opening the door to tailor-made alterations of mutant genes that cause genetic diseases."

"Encouraging news about CRISPR technology surfaced in the American press as I was completing this book," Watson stated, referring to a recent book he published entitled *DNA: The Story of the Genetic Revolution*. Specifically, a piece published online in early August 2017, reported that "scientists have successfully edited the DNA of human embryos to erase a heritable heart condition that is known for causing sudden death in young competitive athletes, opening the doors to a controversial new era in medicine. This is the first time gene editing on human embryos has been conducted in the United States."

"The embryos were allowed to grow for only a few days, and there was never any intention to implant them to create a pregnancy. But scientists also acknowledged that they will continue to move forward with the technology, with the ultimate goal of being able to correct disease-causing genes in embryos."

"This experiment is the latest example of how the laboratory tool known as CRISPR, a type of "molecular scissors," is pushing the boundaries of our ability to manipulate life, and has been received with both excitement and horror."

The horror is born of concern among many about germline editing using technologies presently exemplified by CRISPR technology. In his book *DNA: The Genetic Revolution*, James Watson informs that Francis Collins, a senior scientist in the field of genetics and the director of the National Institutes of Health (NIH) at the time of this writing, and an individual with strong religious beliefs "argues that this is a red line of sorts that should not be crossed. Collins sees no medical need to tinker with 3.5 billion years of evolution, even if the technology were perfectly safe. Collins is quoted as stating: 'Designer babies make great Hollywood babies. They make really bad science and I think they are really bad ethics.' In Collins' view, safety concerns — what some call 'the risk of irreversibility' — remain widespread, as does the lack of a compelling medical need when other tools, such as preimplantation genetic diagnosis, afford parents a degree of choice without meddling irreversibly in the DNA of future generations."

Watson goes on to recount views expressed by another scientist, Eric Lander, who is quoted as stating: "If it is such a good idea, I want to scratch my head and say why didn't evolution try to do that, and increase that in the population?"

Watson reports that a meeting of senior scientists convened to discuss the pros and cons of CRISPR technology concluded that: "there was no compelling reason to sanction human embryo experiments using CRISPR, but nor did the panel recommend a moratorium of CRISPR research. The potential of this technology is too great, too exciting, to cut it off in its tracks and drive it underground."

In the final analysis, Watson concludes that "my view is that, despite the risks, we should give serious consideration to germline gene therapy. Now, having identified many mutations that have caused so much misery over the years, it is in our power to sidestep natural selection. (Meaning let nature take its course and eliminate embryos with genetic diseases that are incompatible with life.)"

Readers interested in the development of this gene technology might want to follow the future of CRISPR.

Watson completes his masterful book with the uncharacteristic following sentiments about genes: "Our DNA, the instruction book of human creation, may well come to rival religious scripture as the keeper of the truth. I may not be religious, but I see much in scripture that is profoundly true. In the first letter of the Corinthians, for example, Paul writes:

> **Though I speak with the tongues of men and of angels, but have not love, I have become sounding brass or a clanging symbol.**
> **And though I have the gift of prophecy, and understand all mysteries and all knowledge, and though I have all Faith, so that I could remove mountains, but have not love, I am nothing.**

Paul has in my judgment proclaimed rightly the essence of our humanity," Watson wrote. "Love, that impulse that promotes our caring for one another, is what has permitted our survival and success on the planet. So fundamental is it to human nature that I am sure that the capacity of love is inscribed in our DNA — a secular Paul would say that love is the greatest gift of our genes to humanity. And if someday these particular genes, too, could be enhanced by our science, to defeat petty hatreds and violence, in what sense would our humanity be diminished?"

11 Mitochondrial DNA

Early in this book, I informed you that the great majority of the DNA in your cells is found in chromosomes in the nuclei of cells. The reason that I did not offer that **all** the DNA in your cells is located in chromosomes stems from the fact that 57 years ago several research teams independently discovered that tiny structures in the cytoplasm of our cells called **mitochondria (Fig. 3-2) also contain DNA (mtDNA)**. A huge surprise! Nuclear and mitochondrial DNA are thought to be of separate evolutionary origin.

Every cell contains hundreds to thousands of mitochondria, which are the seat of biochemical events that generate the energy that all cells require. Mitochondrial DNA contains 37 genes, all of which are essential for mitochondrial functions. Thirteen of these genes provide instructions for making enzymes involved in biochemical reactions that transpire in mitochondria. The remaining genes provide instructions for making transfer RNA (tRNA) and ribosomal RNA, which you recall from the chapter on the genetic code are required for protein synthesis.

You will also recall from the chapter on the repair of damaged DNA that reactive oxygen species (ROS) are a potent source of DNA damage. Regardless of the fact that mitochondrial metabolism generates ROS, mtDNA does not accumulate any more oxidative base damage than nuclear DNA, perhaps because some types of oxidative DNA damage are repaired more efficiently in mitochondrial DNA than in nuclear DNA. Mitochondria have also evolved a unique mechanism that maintains mtDNA integrity by degrading excessively damaged mitochondrial genomes followed by replication of intact repaired mtDNA. The multiple copies of DNA present in mitochondria facilitate this mechanism as templates.

Unlike nuclear DNA, which is inherited from both parents and in which genes are rearranged in the process of recombination during meiosis, there is usually no change in mtDNA from parent to offspring. Because of this, and since the mutation rate of animal mtDNA is higher than that of nuclear DNA, mtDNA is a powerful tool for tracking ancestry through females, and has been used in this role to track the ancestry of many species back hundreds of generations.

During the fertilization of a female egg by a male sperm cell, virtually all the sperm are destroyed. Only the chromosomes found in the head of the sperm are preserved and are used to generate the final version of DNA in the fertilized egg. The sperm mitochondria and its DNA are also broken down. Only the mitochondrial DNA from the egg are used to generate a newly developing fetus. Consequently, **mitochondrial DNA is transferred from mother to daughter, generation after generation after generation** and if you are a woman your mitochondrial DNA is identical to that of your mother and her mother and her mother, etc. The mitochondrial DNA in a son, which he got from his mother, is a dead end. Consequently, sequencing mitochondrial DNA in women is frequently performed to establish familial heritage and in theory should be traceable to a single woman who existed hundreds of thousands of years ago.

Mutations in mitochondrial DNA can lead to a number of illnesses, including exercise intolerance and a disease called Kearns–Sayre syndrome, which causes an afflicted individual to lose full function of heart, eye, and muscle movements. Evidence also suggests that mutations in mitochondrial DNA may be important contributors to the aging process.

12 Ancient DNA

Svante Pääbo, a Swedish biologist and a recognized authority on evolutionary genetics, is one of the founders of the discipline of **paleogentics**, a discipline that deploys the methods of genetics and genomics to study early humans and other ancient populations. Pääbo has been director of the Department of Genetics at the Max Planck Institute for Evolutionary Anthropology in Leipzig, Germany since 1997. In 2014 he published a book entitled *Neanderthal Man: In Search of Lost Genomes*, in which he relates the history of the research efforts in his laboratory.

In his book *A Brief History of Everyone Who Ever Lived*, Adam Rutherford, a British scientist, writer and broadcaster, informs us that, according to traditional paleoanthropology (the study of early humans based on the detailed examination of bones), modern humans (*Homo sapiens*) were believed to have reached Europe about 60,000 years ago, a time when Neanderthals were well established there, albeit in small communities. Examination of Neanderthals' bones also reveals that their anatomy was different from ours. "They were shorter and stockier than us, thick set, barrel chested, with broader noses and chunkier brows." He also points out that "in common parlance they are synonymous with and stigmatized as brutish cavemen, grunting oafs, and 'neantherthal' acts as a byword for lowbrow dimwit thuggery. In fact in the 19th century, the German biologist Ernst Haeckel suggested one skeletal specimen be called *Homo stupidus*.'"

But, Rutherford writes, "nothing indicates they were anything of the sort, nor significantly different from us. They hunted, butchered and cooked large prey." And "there is some evidence that in the last 100,000 years they sewed, made clothes and jewelry and these skills predate the arrival of

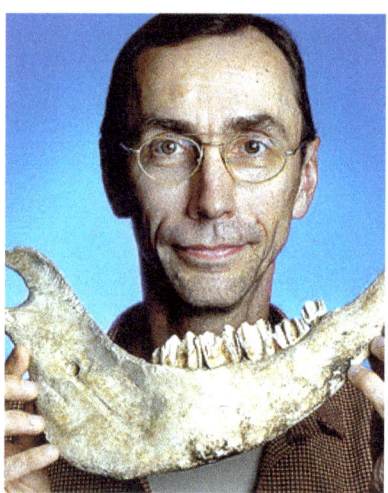

Fig. 12-1. Svante Pääbo.

anatomically modern humans, meaning that the Neanderthals developed these them themselves rather than learning them from the new kids on the block." Rutherford also points out that "Neanderthals had bigger brains than us and though cranial capacity is not something we can specifically correlate with intellectual capabilities, in general, in apes, bigger brains mean more sophistication!" Would it not be sobering to unlock incontrovertible evidence that Neanderthals were in fact intellectually superior to us?

In 1997 Pääbo and his colleagues reported the sequencing of Neanderthal mitochondrial DNA originating from a fossil found in the Neander valley of Germany. The Neanderthal sequence has more in common with modern human mtDNA sequences than with those of chimpanzees, indicating that Neanderthals were undeniably part of the human evolutionary lineage. But there were also significant differences between the Neanderthal sequence and close to a thousand sequences of modern human mtDNA.

Later, mtDNA was recovered and sequenced from two other Neanderthal specimens. The sum of all these efforts led Pääbo to conclude that, though Neanderthals belonged to the evolutionary tree of humans, the Neanderthal branch of that tree was a long distance from the modern human one. These observations suggested that though humans and Neanderthals encountered one another about 40,000 years ago, modern humans eliminated the Neanderthals rather than bred with them.

In subsequent studies, Pääbo and his colleagues sequenced a stretch of **nuclear** DNA from a Neanderthal bone fragment found in a cave in Croatia, that was estimated to be about 40,000 years old and had been kept in a Croatian museum. In 2010 he and his co-authors published an analysis of two million bases — about 50% of the Neanderthal genome, and compared the sequence to those of five modern humans from different parts of the world. The results indicate that, in contrast to us humans whose long-lived ancestors migrated out of Africa to Europe, Neanderthals never lived in Africa. But nuclear DNA sequence comparisons suggest that Neanderthals and humans did in fact interbreed during the period they overlapped, presumably in the Middle East, about 40–75 thousand years ago.

Preliminary studies indicate that 99.7% of the base sequences of the modern human and Neanderthal genomes are identical, compared to humans sharing around 98.8% of sequences with the chimpanzee. Not surprisingly, therefore, it would appear that we are genetically closer to Neanderthals than we are to chimpanzees!

Pääbo's team also reported an extensive genome sequence of a genome extracted from the bone fragment of a female from around 50,000–100,000 years ago found in the Denisova Cave in the Altai Mountains of Siberia. This sequence is neither Neanderthal nor human, indicating the existence of another hominid group called Denisovans, which were apparently more closely related to Neanderthals than to modern humans.

In 2007, mutations in the speech-related gene *FOXP2* identical to those in modern humans were discovered in Neanderthal DNA, suggesting that Neanderthals might have shared some basic language capabilities with modern humans.

13 When and How did DNA Appear on Earth?

When and how did DNA with its large complement of genes arise as the bearers of life? This question has been around for a long time, but there is no definitive answer. The succeeding paragraphs are intended to give the reader a sense of the complexity of this question and some of the thoughts and suggestions that have emerged.

The current notion is that life began with RNA and that in due course RNA-based life switched to DNA because it is better suited to storing information. But there are other opinions about how DNA arose. In an recent article with the title *DNA Could Have Existed Long Before Life Itself*, author Michael Marshall pointed out that "chemists are close to demonstrating that the building blocks of DNA can form spontaneously from chemicals thought to be present on the primordial Earth." If so, that would answer the question about the origin of life on our planet. In 2009, after decades of work, researchers finally managed to generate RNA using chemicals that probably existed on the early Earth. This achievement suggests that RNA might have formed spontaneously — support for the idea that life began in an RNA world and then switched to DNA. Another alternative under consideration is that life may have begun with both RNA and DNA worlds, in which the two types of bases were intermingled and eventually separated. In the final analysis, Marshall quotes Matthew Levy of the Albert Einstein College of Medicine in New York as stating: **"Right now there's nothing to tell us exactly how and when life first used DNA."**

Another burning question concerns life on other planets and asks that assuming that is indeed the case, would it also be DNA-based? In an online essay entitled *Life in the Universe*, Stephen Hawking, the famous English theoretical physicist, cosmologist, author, and Director of Research at the Centre for Theoretical Cosmology at Cambridge University, points out that living beings are generally possessed with both a set of instructions that tell the system how to sustain and reproduce itself (**genes**) and a mechanism to carry out the instructions (**metabolism**). "But," Hawking states, "they don't have to be biological. A computer virus is a program that will make copies of itself in the memory of a computer and will transfer itself to other computers." Hence, in Hawking's view, a computer fits the definition of a living system. Perhaps such "living systems" exist somewhere in the universe?

Hawking concluded his essay by offering the possibility that there are indeed other forms of intelligent life out there, but we have been overlooked. He reminds us that there used to be a project called **SETI**, the **S**earch for **E**xtra-**T**errestrial **I**ntelligence, that involved scanning radio frequencies to determine whether we could pick up signals from alien civilizations. "I thought this project was worth supporting," Hawking related, "though it was cancelled due to a lack of funds. But we should be wary of answering back until we have developed a bit further. Meeting a more advanced civilization at our present stage might be a bit like the original inhabitants of America meeting Columbus. I don't think they were better off for it!"

Fig. 13-1.   Stephen Hawking.

14 The Human Genome Project

The sequence of the bases A, T, C and G in a stretch of DNA can be determined using techniques that are now standard in research laboratories in most if not all academic institutions. However, such laboratories do not have the capacity to handle sequencing the genome of organisms such as humans, which is estimated to contain about 25,000 genes as well as many more bases that are not part of any gene.

Somewhere around the decade of the 1980s, scientists began discussions about establishing a project to sequence the entire human genome. After much discussion and innumerable meetings, two government agencies in the U.S., the Department of Energy and the National Institutes of Health, developed a memorandum of understanding in order to coordinate plans and set the clock for the initiation of the Project in 1990. The $3-billion project was formally initiated in 1990 by the U.S. Department of Energy and the National Institutes of Health and was expected to take 15 years. In addition to the United States, geneticists in the United Kingdom, France, Australia, China and myriad other spontaneous relationships joined the Human Genome Project **(HGP)**.

Widespread international cooperation and advances in the field of DNA sequence analysis as well as major advances in computing technology facilitated "a rough draff" of the human genome in 2000 that was announced on TV jointly by U.S. President Bill Clinton and British Prime Minister Tony Blair on June 26, of that year. Announcement of the essentially complete DNA sequence of the human genome was made on April 14, 2003, two years earlier than initially planned. The HGP was one the most expensive undertakings in biology. The sequence of the DNA is stored in databases available to anyone on the **Internet**. The U.S. National Center for Biotechnology Information and sister organizations in Europe and Japan, house the gene sequence in a database known as *GenBank*. The genome published by the HGP does not represent the sequence of every individual's genome. It is the combined assortment of a small number of anonymous donors, all of European origin.

In 1998, a similar, privately funded undertaking was launched by an American researcher Craig Venter and his firm *Celera Genomics*. Venter was a scientist at the NIH during the early 1990s when the HGP was initiated. Work on the interpretation and analysis of data already accumulated is still in its initial stages.

It is anticipated that detailed knowledge of the human genome will provide new avenues for advances in medicine and biotechnology. But practical results of the HGP emerged even before the work was finished. For example, a number of private companies began offering simple ways of administering genetic tests that can show predisposition to a variety of illnesses, including breast cancer, cystic fibrosis, liver diseases and others. Additionally, it is anticipated that the causes of cancers, Alzheimer's disease and other areas of clinical interest will benefit from genome information and may lead to significant advances in their management.

The analysis of similarities between DNA sequences from different organisms is also opening new avenues in the study of evolution. Many questions about the similarities and differences between humans and our closest relatives (apes and monkeys) and other mammals are expected to be answered in this project.

The Human Genome Project, a massive international research effort to sequence and map all the genes of members of our species, *Homo sapiens*, is one of the great feats of exploration in history.

15 Conclusion

I hope that you enjoyed reading this book and did not experience too much difficulty understanding most, if not all of its content; and more importantly, that you learned something about your genes and how they function.

At various places in the book I mentioned, quoted and acknowledged Siddhartha Mukherjee, author of **The Gene — An Intimate History**. A chapter in Mukherjee's book entitled, **The Book of Man (in Twenty-Three Volumes)** (borrowed from our twenty three pairs of chromosomes) presents a summation of the human genome that bears repetition as the final words of **Learning About Your Genes**.

"It has 3,088,286,401 letters of DNA; a little over three billion (give or take a few).

It is divided into 23 pairs of chromosomes — 46 in all — in most cells in the body. All other apes, including gorillas, chimpanzees, and orangutans have 24 pairs. In one point in hominid evolution, two medium-size chromosomes in some ancestral ape fused to form one.

It encodes about 20,687 genes in total — only 1,796 more than worms, 12,000 fewer than corn, and 25,000 fewer genes than rice or wheat.

It is fiercely inventive. It squeezes complexity out of simplicity. It orchestrates the activation or repression of certain genes in only certain cells and at certain times, creating unique partners for each gene in time and space, and thus produces near-infinite functional variation out of its limited repertoire.

It is dynamic. In some cells it reshuffles its own sequence to make novel variants of itself. Parts of it are surprisingly beautiful. On a vast stretch of chromosome 11, for instance, there is a causeway dedicated entirely to the sensation of smell.

Genes, oddly, comprise only a miniscule fraction of it (meaning the genome). An enormous proportion — a bewildering 98% — is not dedicated to genes *per se*, but to enormous stretches of DNA that are interspersed between genes or within genes. These long stretches encode no RNA and no protein: they exist in the genome either because they regulate gene expression, or for other reasons, or for reasons that we do not yet understand, or for no reason whatsoever (i.e., they are "junk" DNA).

It is encrusted with history. Embedded within it are peculiar fragments of DNA — some derived from ancient viruses — that were inserted into the genome in the distant past and have been carried passively since then.

It has repeated elements that appear frequently. A pesky, mysterious 300-base-pair sequence called *Alu* appears and reappears tens of thousands of times, although its origin, function, or significance is unknown.

It has enormous "gene families" — genes that resemble each other and perform similar functions — which often cluster together.

It contains thousands of "pseudogenes" — genes that were once functional but have become non-functional, i.e., they give rise to no protein or RNA.

It accommodates enough variation to make each one of us distinct, yet enough consistency to make each member of our species profoundly different from chimpanzees and bonobos, whose genomes are 96% identical to ours.

Although we fully understand the genetic code — i.e., how the information in a single gene is used to build a protein — we comprehend virtually nothing of the *genomic code* — i.e., how multiple genes spread across the human genome coordinate gene expression in space and time to build, maintain, and repair a human organism.

> "It is inscrutable, vulnerable, resilient, adaptable, repetitive, and unique.
> It is poised to evolve. It is littered with the debris of its past.
> It is designed to survive.
> It resembles us."

* * *

Siddhartha Mukherjee is an adept writer and ***THE GENE — AN INTIMATE HISTORY*** is the definitive non-fiction work about genes. Some of you may be prompted to read this book. If you are, don't be surprised to discover that though clearly distinct from a textbook, Mukherjee's tomb of 568 pages is considerably more advanced than the one you have just read — which was one of my principal motivations to write this book.

Glossary

A

Acquired mutations   mutations that are not heritable

Adenine   one of the four bases in DNA and RNA

Alternative splicing   alternative splicing, or differential splicing, is a regulated process during gene expression that results in a single gene coding for multiple proteins

Amino acids   any of a large number of compounds found in living cells, that contain carbon, oxygen, hydrogen, and nitrogen, and join together to form proteins. *Amino acids* contain a basic *amino* group (NH_2) and an acidic carboxyl group (COOH), both attached to the same carbon atom

Alzheimer's Disease   a progressive mental deterioration that can occur in middle or old age due to generalized degeneration of the brain

Avery, Oswald Theodore   a distinguished Canadian-born bacteriologist and research physician who discovered that genes are made of DNA

B

Bacteriophage   a virus that parasitizes a bacterium by infecting and reproducing inside it

Base a chemical constituent of DNA and RNA, of which there are four: adenine, cytosine, thymine and guanine. RNA contains the base uracil instead of thymine

Base pair two bases that face each other in a double-stranded DNA molecule

Berg, Paul a scientist at Stanford University

Boyer, Herbert a scientist at the University of California in San Francisco

Brenner, Sydney a South African-born molecular biologist who was instrumental in helping decipher the genetic code

Buckland, Richard a suspect in a series of murders in Leicester, England who was released and is the first accused criminal to be found not guilty of a crime based on DNA fingerprinting

C

Celera Genomics a private biotechnology company founded by Craig Venter in 1998

Chase, Martha a deceased scientist who, together with Alfred Mirsky, helped prove that genes are made of DNA

Chromosomes thin thread-like structures in the nuclei of cells, where DNA resides

Cobb, Matthew a British author who wrote a book on the history of genomics and genetics entitled *Life's Greatest Secrets*

Codon a term that describes three successive bases in DNA

Crick, Francis a deceased famous British molecular biologist who, together with James Watson, solved the structure of DNA in 1953

CFTR gene | a gene that is mutated in the genetic disease cystic fibrosis

Cystic fibrosis (CF) | a genetic disorder that affects the respiratory and digestive systems and the sweat glands, and is often fatal

Cytosine | one of the four bases in DNA

Cytoplasm | the liquid interior of a cell in which all intracellular structures reside

D

Denisovans | a recently discovered branch of extinct humans that may have once interbred with us

Demerec, Milislav | a Croatian-born scientist who served as director of the Cold Spring Harbor Laboratory from 1941 to 1960

DNA | the hereditary material in humans and almost all other organisms

DNA fingerprinting | a laboratory technique used to establish a link between biological evidence and a suspect in a criminal investigation

DNA polymerase | an enzyme that manufactures new DNA

DNA repair | a process that repairs damaged or incorrect bases in DNA

DNA replication | a process by which a DNA molecule is copied to generate new DNA

DNA sequencing | a technique whereby the sequence of bases in a DNA strand is obtained

DNA splicing | the removal of introns (intervening sequences) from the primary transcript of a discontinuous gene during the process of transcription

Dulbecco, Renato | an Italian-born molecular biologist who was one of two scientists who independently discovered a DNA repair process called photoreactivation

E

Enzyme	a protein that helps speed up a biochemical reaction
Excision repair	a process whereby damaged or incorrect bases are removed from DNA
Exon	the coding region of a gene, that also contains introns

F

Fleming, Alexander	the person who discovered the first antibiotic, penicillin
Franklin, Rosalind	a British scientist who worked on deciphering the structure of DNA

G

Gene	a unit of heredity that is transferred from parent to offspring and is held to determine some characteristic of the offspring
Gene cloning	a process in which a gene of interest is located and copied (cloned) from DNA extracted from an organism
Gene splicing	a process that involves cutting out part of the DNA in a gene and adding new DNA in its place
Gene therapy	the transplantation of normal genes into cells in place of missing or defective ones
Genetic code	the set of rules by which information encoded within genetic material (DNA or mRNA sequences) is translated into proteins by living cells
Genetic disorders	genetic problems caused by one or more abnormalities in the genome, especially condition that is present from birth (congenital)

Genetic mapping	genetic map distances based on genetic linkage information
Genetics	the study of heredity and the variation of inherited characteristics
Genome	a word that describes all the genes in a given organism or individual
Genomics	the branch of molecular biology concerned with the structure, function, evolution, and mapping of genomes
Genotype	the genetic makeup of an organism or group of organisms with reference to a single trait, set of traits, or an entire complex of traits
Germiston	a city in South Africa where Sydney Brenner was born
Griffith, Frederick	a British scientist who discovered DNA without realizing it
Guanine	one of four bases in DNA

H

hereditary mutations	mutations that are inherited from one or more parents
Hershey, Alfred	an American scientist who worked with Martha Chase
Homo sapiens	Latin for humans
Human Genome	the complete set of nucleic acid sequence for humans (*Homo sapiens*), encoded as DNA within the 23 chromosome pairs in cell nuclei
Human Genome Project	an international scientific research project with the goal of determining the sequence of nucleotide base pairs that make up human DNA, and of identifying and mapping all of the genes of the human genome from both a physical and a functional standpoint

Huntington's disease   a hereditary disease marked by degeneration of brain cells and causing chorea and progressive dementia

I

Intercalate   insert something between layers in a crystal lattice, geological formation, or other structure such as DNA

Intron   a segment of a DNA or RNA molecule that does not code for proteins and interrupts the sequence of genes

J

Jaenisch, Rudolf   a German scientist

Jeffreys, Alec   a British geneticist, who developed techniques for DNA fingerprinting and DNA profiling which are now used worldwide in forensic science to assist police detective work and to resolve paternity and immigration disputes

Johannsen, Wilhelm   a Danish botanist, plant physiologist, and geneticist

K

Kearns–Sayre syndrome   a rare neuromuscular disorder

Kelly, Ian   a suspect in a murder case in Leicester, England, investigated by Alec Jeffreys

Kelner, Albert   an American scientist who discovered enzymatic photoreactivation

M

Mann, Linda   a woman found murdered in Leicester

Marshall, Michael   author of *DNA Could Have Existed Long Before Life Itself*

Mendel, Gregor Johann	a friar at a monastery in Czechoslovakia
Meiosis	a type of cell division that results in four daughter cells, each with half the number of chromosomes of the parent cell, as in the production of germ cells
Meselson, Matthew	an American scientist, formerly at Harvard University
Messenger RNA (mRNA)	the form of RNA in which genetic information transcribed from DNA as a sequence of bases is transferred to a ribosome
Metabolism	the chemical processes that occur within a living organism in order to maintain life
Minisatellites	any of numerous DNA segments located mainly near the ends of chromosomes, that consist of repeating sequences of at least 5 but usually not more than 100 nucleotides and that are useful in DNA fingerprinting
Meiosis	a specialized type of cell division that reduces the chromosome number by half, creating four haploid cells, each genetically distinct from the parent cell that gave rise to them
Mismatch repair	a DNA repair process that corrects mismatched bases in DNA
Mitochondria	rod-shaped organelles that can be considered the power generators of the cell, converting oxygen and nutrients into adenosine triphosphate (ATP), the chemical energy "currency" of the cell that powers the cell's metabolic activities
Mitochondrial DNA	an extranuclear double-stranded DNA found exclusively in mitochondria, that in most eukaryotes is a circular molecule and is maternally inherited

Mitosis — a type of cell division that results in two daughter cells each having the same number and kind of chromosomes as the parent nucleus

Model organism — a non-human species that is extensively studied to understand particular biological phenomena, with the expectation that discoveries made in the organism model will provide insight into the workings of other organisms

Mukherjee, Siddhartha — an Indian-American physician, oncologist, and author best known for his 2010 book *The Emperor of All Maladies*

Mutation — the changing of the structure of a gene, resulting in a variant form that may be transmitted to subsequent generations, caused by the alteration of single base units in DNA, or the deletion, insertion, or rearrangement of larger sections of genes or chromosomes

N

Neanderthals — a species distinct from modern humans that became extinct because of climate change or interaction with modern humans and was replaced by modern humans moving into their habitat between 45,000 and 40,000 years ago

NIH — the National Institute of Health located in Bethesda, Maryland, U.S.A.

Nucleotide — one of the structural components, or building blocks, of DNA and RNA. A nucleotide consists of a base (one of four chemicals: adenine, thymine, guanine, and cytosine) plus a molecule of sugar and one of phosphoric acid

Nirenberg, Marshall   co-recipient, with Robert William Holley and Har Gobind Khorana, of the 1968 Nobel Prize for Physiology or Medicine. He was cited for his role in deciphering the genetic code

Nuclear DNA   the genome located in the nuclei of cells

O

Olfactory receptors   expressed in the cell membranes of olfactory receptor neurons and are responsible for the detection of odorants (i.e., compounds that have an odor) which give rise to the sense of smell

Oxford University   a collegiate research university located in Oxford, England

P

Paleogenetics   the study of the past through the examination of preserved genetic material from the remains of ancient organisms

Pääbo, Svante   a Swedish biologist specializing in evolutionary genetics. One of the founders of paleogenetics, he worked extensively on the Neanderthal genome

Penicillin   an antibiotic

Phenotype   the observable physical properties of an organism, including the organism's appearance, development, and behavior

Physical gene maps   physical gene maps display the physical distance between genes

Phosphorus   the second most abundant mineral in one's body

Photoreactivating enzyme   an enzyme that repairs pyrimidine dimers in DNA

Photoreactivation   the process that restores pyrimidine dimers in DNA to their normal monomeric state by an enzyme called photoreactivating enzyme

Pitchfork, Colin   a convicted murderer who killed two women in Leicester, England

Plasma membrane   a membrane that surrounds the outside of a cell

Plasmid   a circular DNA that infects bacteria that is much used in the laboratory manipulation of genes

Polypeptide   a linear organic polymer consisting of a large number of amino acid residues bonded together in a chain, forming part of (or the whole of) a protein molecule

Proflavine   a chemical that was used experimentally to intercalate between 2 base pairs in DNA

Protein   any of a class of nitrogenous organic compounds that consist of large molecules composed of one or more long chains of amino acids, and are an essential part of all living organisms, especially as structural components of body tissues such as muscle, hair, collagen, etc., and as enzymes and antibodies

Pseudogene   a segment of DNA that is related to a real gene and that has some functionality relative to that gene

Pyrimidines   the bases thymine (T) and cytosine (C) present in DNA and uracil (U) and cytosine (C) present in RNA

R

Reactive oxygen species (ROS)   chemically reactive chemical species containing oxygen. Examples include peroxides, superoxide, hydroxyl radicals, and singlet oxygen. ROS are highly reactive with DNA

Recessive relating to or denoting heritable characteristics controlled by genes that are expressed in offspring only when inherited from both parents, i.e., when not masked by a dominant characteristic inherited from one parent

Recombinant DNA technology joining together of DNA molecules from two different species that are inserted into a host organism to produce new genetic combinations that are of value to science, medicine, agriculture, and industry

Restriction enzyme an enzyme produced chiefly by certain bacteria, having the property of cleaving DNA molecules at or near a specific sequence of bases

Restriction fragment length polymorphism (RFLP) a difference in homologous DNA sequences detected by the presence of fragments of different lengths after digestion of DNA with specific restriction endonucleases

Ribosome a minute particle consisting of RNA and associated proteins, found in large numbers in the cytoplasm of living cells. It binds messenger RNA and transfer RNA to synthesize polypeptides and proteins

Ribosomal RNA ribosomal ribonucleic acid (rRNA) is the RNA component of the ribosome, and is essential for protein synthesis in all living organisms

Ribonucleic acid (RNA) a nucleic acid that is generally single stranded (double stranded in some viruses) and composed of repeating nucleotide units of the sugar ribose, phosphate groups, and

	nitrogenous bases. A polymeric molecule essential in various biological roles in coding, decoding, regulation, and expression of genes
RNA polymerase	RNA polymerase (or, more fully, ribonucleic acid polymerase, abbreviated RNAP or RNA pol), also known as DNA-dependent RNA polymerase, is an enzyme that produces primary transcript RNA in cells. RNAP is necessary for constructing RNA chains using DNA genes as templates, a process called transcription

S

Salk Institute	The Salk Institute for Biological Studies is an independent, non-profit, scientific research institute located in La Jolla, San Diego, California, United States. It was founded in 1960 by Jonas Salk, the developer of the polio vaccine
SETI (Search For Extra-terrestrial Intelligence)	a not-for-profit research organization, the mission of which is to explore, understand and explain the origin and nature of life in the universe
Semen	a viscous whitish secretion of the male reproductive organs, containing spermatozoa and consisting of secretions of the testes, seminal vesicles, prostate, and bulbourethral glands
Silent mutations	mutations that have no obvious effect
Spliceosome	a cellular structure that removes introns from a transcribed pre-mRNA, a type of primary transcript during the process of transcription

Staphylococcus a type of bacteria. There are over 30 types, but *Staphylococcus aureus* causes most "staph" infections

T

The Double Helix James Watson and Francis Crick collected and interpreted key evidence to determine that DNA molecules take the shape of a twisted ladder — a *double helix*

Thymine one of the four bases in DNA

Transcription the process by which the information in a strand of DNA is copied into a new molecule of messenger RNA (mRNA)

Transformation a thorough or dramatic change in function or appearance

Transforming Principle a groundbreaking experiment, performed in 1928 by Frederick Griffith that established that there was a transforming principle in bacterial extracts that changes their function

Transfer RNA small RNA molecules that carry amino acids to the ribosome for polymerization into a protein

Translation During translation amino acids are inserted into the growing polypeptide (protein) chains when the anticodon of the transfer RNA pairs with a codon on the mRNA being translated; a step in protein biosynthesis wherein the genetic code carried by mRNA is decoded to produce the specific sequence of amino acids in a polypeptide chain. Translation follows transcription

Triplet code the standard version of the genetic code, in which a sequence of three nucleotides on a DNA or RNA molecule codes for a specific amino acid during protein synthesis

U

Uracil

a pyrimidine base that is one of the fundamental components of RNA, in which it forms base pairs with adenine (A)

V

Venter, Craig

founder, chairman, and CEO of the J. *Craig Venter* Institute, a not-for-profit research organization with about 250 scientists and staff dedicated to human, microbial, plant, synthetic and environmental genomic research

W

Wambaugh, Joseph

author of the book *The Blooding,*

Watson, James

an American molecular biologist, geneticist and zoologist, best known as one of the co-discoverers of the structure of DNA in 1953

Wilkins, Maurice

a New Zealand-born British physicist and molecular biologist and Nobel laureate whose research contributed to the understanding of X-ray diffraction

X

Xeroderma Pigmentosum

a genetic disease resulting in defective DNA repair

X-ray crystallography

a technique used for determining the atomic and molecular structure of a crystal, in which the crystalline atoms cause a beam of incident X-rays to diffract into many specific directions

Index

Lightning Source UK Ltd.
Milton Keynes UK
UKHW020616280220
359470UK00005B/46